ISBN 978-0-359-87837-6

This book is dedicated to my wife Nina

who made it possible.

Cold hearted orb that rules the night
Removes the colors from our sight
Red is gray and yellow white
But we decide which is right
And which is an illusion

The Moody Blues

CONTENTS

Introduction

This is my third book. Previous to this I had two works which were basically commentaries on current news events mostly in the U.S. but sometimes international. They are titled, The Mild Equator and Elephants in Real-Time. My commentaries mostly focused on the illogical absurdities of human behavior and wayward attitudes taken in everyday situations. I found gross illogic in "normal" organizational and business issues centered around something not quite as obvious. For example a convicted serial killer gets five life sentences rather than six because he discloses the location or names of additional victims. The obsession with social media, even by adults or political correctness rather than candor also appeared as odd to me. Why would reality as we know it be blatantly adulterated with illogical premise and be accepted as rain?

This prodded me to wonder why people were doing what they were doing and thinking along the lines of nonsense. It was as if there was something circumventing basic logic and reasoning. I decided to create a perspective of human behavior and thought starting from square-one, the reality of existence.

That reality has at its core the science of energy and being. This is the very crux of human reality and human reality is the matrix from which all other human issues grow.

Although the book deals with physics it is not an omnibus of mathematical squiggles nor a composite of scientific words requiring an academic text. If you have had high school level physics way back when it's probably enough to understand what's going on.

On a similar note the complexities of theoretical physics might be mentioned but not considered at any meaningful depth. Theoretical physics is largely structured around second and third order probabilities which means a probable conclusion is based on a probable premise which has as its support another probable premise. The concept of chance causation however is not entirely dismissed and is used in the most logical way. For example it might be posited that a tree falling in the woods cannot be absolutely assumed to make a sound if it falls because there are no witnesses to the event. This is absolutely correct but for the sake of sanity and as far as the book is concerned the tree will be considered to create a sound because a thousand other instances which were witnessed evidenced a sound on the tree falling. It might sound ridiculous but hours of professional contention in the theater of theory has been spent on this or a similar debate.

On a still similar note the phrase pure physics was chosen over the conventional metaphysics for a number of reasons. If you Google metaphysics you will find dozens of metaphysical systems and philosophers all tagged to that word. A good deal of this book is hit upon by someone else or some other system of reasoning over the past 5,000 years, but then again so is every other topic under the sun when it comes to metaphysics. Metaphysics also tries to draw parallels between things like the human soul and moral judgment. It tries to define a development of human reason based on man's understanding of man. It posits the transformation of man's brutality, such as the accepted use of crucifixion and torture, to humane treatment was a development of intellect and morality.

In reality those brutalities existed on the basis of necessity because there was no universal way to convey a message to a governed group of people. That's where the book comes in; long before man could think about what he did he did it for a reason.

Much of the time in conventional metaphysics there is a vague reference to things like energy or a mysterious interconnection between matter and energy but there are also allusions to a spirit world and a particular deity of influence.

In this assessment based on pure physics the subject is confined to actual, factual science and proven or previously documented, accepted normal human behavior and thought.

As far as human reality is concerned the world is in fact as most of the human population sees it. It's everyday life and history in major and minor events and an anticipated future or direction based on that history. The idea that physical reality is as determined by the laws of physics is a proven truism. Human reality as perceived and actually lived by beggars and kings is something which no one can deny exists and presents two-sides of a real coin.

If you consider the sun shines alike on beggars and kings, the sun is an element of pure physics and kings and beggars are under the umbrella of human reality as life in the real world. Beggars and kings however are created by man and the sun is not. The sun was also there first. So what's the big deal? The big deal is the question of how did humanity evolve into its own interpretation of reality and culture when the very essence of all governing power is a translation of energy thru the laws of physics? Stated more specifically; the core of all physics based definitions is the concept of energy.

There is only one acronym used throughout the book and that is COE representing Conglomeration Of Energy or Consolidation Of Energy or even Composite Of Energy. Everything is comprised of energy so when I refer to humans I will call out a human COE or non-human COE or living COE or non-living COE. If everything is built out of energy, even solid matter, reality itself is a product of how that energy does its thing. The difficult part to wrap your head around is how human consciousness and values came to be justified as more than building blocks of pure energy. This is what the book attempts to explain. Everything from thoughts and ideas to memory to even fundamentals as words are based in energy.

A number of other terms are also referenced frequently throughout the book. One set of terms is imaginary plane vs. non-imaginary plane. This refers to things summoned in thought as opposed to things which are tangibly real. Metaphysics harps on this issue but I keep it simple rather than going deeper as to the nature of thought and its reflection of man's soul. Another term is energy construct which is self-explanatory; something constructed of energy. Another is ambient energy matrix which means the sea of intermingling energies which might be atmosphere or simply everything else that does not have a defined shape or property. Still another set of related terms concerns the transformation of energy into different forms. The terms energy transition and energy transposition or the transposing of energy are used interchangeably. They refer to this process as this is the basis of all reality including life and time.

Many times the book will refer to word origins and basic dictionary definitions to understand their significance and role in the dynamic process of human life. The reason for this is that words and language are the energy laden symbols of more constructs of energy both man and nature has assembled. How those word symbols were created way back when is a clue as to how energy was interpreted before civilization had a chance to adulterate the meaning to its own advantage or by way of mutation based on modern consciousness. At face value perception is a function of energy and the translation of that energy into a usable, storable symbol that can be manipulated fostered the creation of words by necessity. Necessity is not only the mother of invention but it is a no-frills demand which does not mince words.

As far as philosophy is concerned the book transcends the ideas of the historical greats although it does discuss traditional philosophical subjects like justice, politics and right and wrong. The difference between then and now is the advancement of science and physics and the ability to focus on one subject. In days of old they were still sorting out the differences between concepts like justice and religion.

Even though the ancient Greeks talked about things like atoms they did not have the practical grasp of energy in all forms and it's transitional properties. They could not consider it as the largest controlling factor for the creation of human reality. That void in knowledge lead the early philosophers down a road of assumptive logic based on human behavior. Reality was interpreted as a function of man and his natural strengths and weaknesses as well as an awareness of his place in civilization.

Central to much early philosophies was the concept of utopia. Utopias were created on an imaginary plane. Those civilized versions of utopia could only be achieved by man's awareness of his own behavior and the determination to mold that behavior in a certain direction. So at this point in time the object of life, its origins and the mechanics of its function became a reality of direction and instruction to achieve that imaginary utopia thru organization.

Philosophy was in effect enlisted as a roadmap to be followed towards an aspired way of life by defining and achieving that organization. It was closer to our political system of today rather and not the world or life as we know it. Following the same argument as the structure of the grain of sand with respect to the building to understand the big picture later let's say the grain of sand existed first and the huge building was an outgrowth of that grain of sand. Confused yet?

As Einstein and others have proven everything is made up of energy. Physical matter is compacted, intertwined factions of energy. An atom of carbon is a *conglomeration of energy.* Diamonds are compacted and sintered carbon in a crystalline form of a specifically structured, conglomeration of energy. Gold is a conglomeration of energy that was randomly dispersed in minute amounts when the planets were formed. The moon is a conglomeration of energy. *(*For simplicity's sake I will refer to a *conglomeration of energy* or *consolidation of energy* or *composite of energy* from here on as a *COE.)* A physicist will once again pick apart that carbon atom, split it off or adulterate it in an effort to find significance for it in the real world or how it might determine the value of coal as fuel.

The big picture in pure physics recognizes that the carbon atom is indeed just a COE like any other COE. The arrangement of energy in the carbon atom is immaterial to the core concept of pure physics in the light of reality. The conventional physicist trying to discern the difference between a carbon atom and an atom of iron is like studying the difference between a round man-hole cover and hexagonal man- hole cover. There is a difference but is it really significant as far as understanding the purpose of a sewer system?

The key to understanding the difference in anything within the universe as we know it (once again because we can only go back so far) lies in the differences in the arrangements of energy.

In conventional physics *differences in energy types* are a major player. Chemical energy is contrasted with heat energy or electrical energy or nuclear energy or kinetic energy, and so on. Comparing these forms of energy is like comparing apples and oranges when in fact they are all fruit. The real core issue when talking about energy and the genuine parameters which determine everything in the universe with respect to energy comes down to two things: *arrangement and quantity.* Every COE can be defined by it's internal arrangement and quantity. But it isn't so much the different class of energy or energy state as the attractive forces which comprise energy at its core. Out in the vastness of the beginning, or imagined end for that matter, pools of attractive forces are in constant flux much like whirlpools and ocean currents.

The only thing that exists are forces without permanent physical shape which we call energy. As they intertwine in different quantities and come to temporal rest in particular arrangements they achieve the labels we give them based on our perception of those energies. Be it wind or rain or an atom of carbon it's still an arrangement of a quantity of energy, a COE glued at random thru sub-atomic forces.

Words and Definitions: ENERGY: The human definition for energy is strength, power vitality and the capacity to do work. *Work* is defined as a required process to do something.

So energy as understood by humanity is something that will perceptibly bring about a change to something else. It could be a position in space, a change in size, a change in temperature or a host of others capabilities. The unique property about this is that the thing that is going to change is also energy; a conglomeration of energy, a COE. What we are defining in a pan-universal sense is that energy is continually acting on energy. The only difference is the resulting quantity and arrangement which we see at any point in time

Even the ancient Greeks recognized a movement of energy thru things like a waterfall, an ox or a hammer to change some other form of matter or COE. If that COE was changed correctly into another COE another proposed transference of energy could be enacted, like the planting of crops. The crops are then further energy-processed into food for the sustaining of biological life, their life.

Words and Definitions: LIFE: the condition that distinguishes animals and plants from inorganic matter, including the capacity for growth, reproduction, functional activity, and continual change preceding death.

As with many definitions the chicken or the egg condition often prevails. Life is something which is not death. When push-comes-to-shove, the conditions of life in common plant mold is not really different from iron oxidation (rust). It's a COE which is acted upon by other COEs and ultimate acts upon other COEs in close proximity.

For the sake of argument, the human interpretation of life occurred about 4 billion years ago. Some kind of blue-green algae was developed out of other chemical elements. The reality is COEs were acting on each other in a haphazard way with no rhyme or reason other than accumulated, random, attraction the same as they were from the beginning of time whenever that was. Eventually the push-and-pull dynamic created a quantity of energy in a particular arrangement which would replicate itself rather than change into a perceivably different form. Hypothetically if the COE was water it would not evaporate into gasses it would produce more water by extracting elements out of the environment. It was not water in this case but some other COE called *organic*.

As the organic series of replications were acted upon by other environmental COEs the design changed into a multitude of forms with all still retaining the capacity for replication. One of the form design changes happened to be what we label as human.

Even though the subject of evolutionary development can be debated, for simplicity's sake the book will consider it valid. Even the concept of what is living and what is not can be debated but that's another story. There is also a subjective reality for all other living things even though we like to think our reality is the only reality. That reality, or those realities have been there since day one, whenever that was.

Words and Definitions: REALITY : It's another word that is defined by another word, existence; and existence is defined by reality. We also consider reality to be defined by "this world" or "of this world." This is also less than perfect because planets and beyond are not of this world but what we consider real because we are of this world and can perceive them. Confused yet?

As far as the beginning of the universe, the Big Bang is the closest we can come to a half-theoretical reason to believe something existed before the earth and planets. This is something which is common knowledge but can't be proven.

Strange as it may seem there is more weight to the argument that COEs were always the primary building blocks of reality and there was no predecessor that can be defined. The fact that COEs will also be the non-definable end to reality really gives support for the religious contention that God, a non-definable entity, had no beginning and no end. Most religions maintain this postulate and it indeed may be inferring the same premise. So here we are COEs with no traceable beginning other than a definition which man created and called life.

4

No visible means of support

and you have not seen nothin' yet

Everything's stuck together

I don't know what you expect staring into the TV set

Talking Heads

Chapter Two

The Physical Picture

Here are some aspects of conventional and not so conventional physics presented as a back-drop for the understanding of reality.

Mass, Gravity, Magnetism, Attraction, Etc.

In conventional physics the forces that cause molecules to bond and gravity and magnetism are considered different forces. This bit of micro-analyzing is the equivalent of requesting a chemical-spectral analysis of paint you want to purchase to paint a yard shed. As far as human reality is concerned the same forces that cause tree sap to cling to a car hood are the same forces that keep a space craft at a steady descent on the moon which are the same forces that keep an egg shell together around an egg.

Physicists theorize gravity is something altogether unique. They consider it a function of something called space-time but for all intents and purposes it is the function of a COE intertwining with many other COEs. Granted they may be extremely large COEs but COEs none-the-less. By virtue of the core building blocks of energy there exists some variety of attractive forces that to date cannot be defined although they are agreed to exist. Those swirls of force maintain an active potential in sufficient quantity and arrangement to cause the conditions known as gravity, bonding or magnetism. You need to remember these are man-made labels for forces that would exist anyway even if man himself was never created.

The energies in question might be defined as chemical, electrical, light or even mechanical but at the end of the day it's all energy creating the effect we call force or attraction. The fact that we do not have an explanation for this renders a diverse definition by classification as plausible but none-the-less not on the exact mark.

The closest we can come today to an explanation is acknowledging some condition which exists even in a vacuum. Very generally speaking some kind of *presence* is contrasted with a *non-presence*. The presence infinitely invades the non-presence because there is nothing to stop it. To take it one step further the non-presence may in fact just be a lesser presence. A lesser presence contrasted with a greater presence is the equivalent of a non-presence as contrasted with an absolute presence. The net effect is a *polarization* which results in a complete relationship interaction called energy.

The accumulation of energy or energies in any one of a multitude of arrangements is responsible for all humanly detectable things and forces, including gravity, magnetism and the capabilities of gorilla glue.

In conventional physics the idea that light bends thru gravity because gravity is warped thru space may be true but it's more plausible that photons of light are *influenced* by gravitational forces because it is a form of energy like everything else. Because the weight of a photon has not been established does not mean it does not exist nor that it is not influenced by gravity. The logic behind the theory is that mass and gravity are contingent upon each other.

Because a detectable mass in photons has not been established and gravity is effective based on mass does not mean photons may be paralleled by another force which impacts trajectory. This may explain the bending. If the unknown force is exclusively evident in the presence of gravity, gravity might be posited to cause the light bending until that unknown force is discovered which is where we are at this point in discovery.

Electrons which are influenced by electromagnetic forces, as in a cathode-ray tube, do have mass. Also, since light energy is energy and mass is comprised of energy, photons probably have some mass although undetectable. Radio waves also do not have mass by today's standards. The shortest, most probable explanation is that all energy is influenced by other energy and in some cases the attractive forces are so small they are undetectable. Even the sun, a swirling ball of gas, heat and light does exert gravitational forces in space.

On a simpler note the mechanical lock on a front door, a nail holding two pieces of wood together or a screw-and-nut are all able to do the task they do by virtue of those same forces of energy. From the beginning of time energy has been uniquely arranged, by natural events or humanity into contrasting media which by virtue of competing density "locks" other COEs together like a Chinese puzzle. The opposing forces are oriented to limit the transposition of energy within a physical space. The lack of free movement was purposely or by nature arranged achieving that end condition.

The lock on the door or nail in the wood pieces are considered mechanical bonds rather than molecular bonds because of the higher-order arrangement of complete molecules which make up the materials involved. The molecules in the lock are bonded within the component parts but the parts are blocked from an "open" condition because of component density and molecular bond strength within the opposing component parts. In the case of the wooden pieces the molecules are bonded in the steel in a nail but do not bond to the wood molecules. They are forced to intermesh by surface pressure created by attractive forces in the molecules of wood.

Building Blocks Of All That Ever Was
Light, Heat, X-Rays, Radio Waves, Electricity, Etc.

In 1864 Heinrich Hertz discovered passing a high voltage current thru lengths of wires caused a spark to occur between a gap in other lengths of wires which were not physically connected to the first set of wires. This was the first proof of an energy transmission thru space that could not be perceived by a human being in natural form. Although the understood value of this as radio technology came about 50 years later, the real value as far as an understanding of reality came with the initial discovery. Hertz's discovery proved there were other forms of energy operating undercover perhaps since the earth was created or even before.

Shortly after in 1894 X-Rays were discovered as another form of energy which could not only pass thru air but solid matter as well. Radio waves can also do this but the demonstrated effect of radio waves was not as dramatic as actually seeing the effect as evidenced in photographic plates.

Similar to radio, X- Ray technology later exploited the waves for medicine and beyond but the gem of the discovery was the implied size range of rogue energies. Essentially molecular constructs of energy in the form of matter are so relatively large in relation to other forms of "free" energy that one can pass thru the other's locking field matrix without being disturbed nor disturbing the field matrix itself. This presented a perspective into the range of size between different quantities and arrangements of energy. The conclusion being the fundamental building blocks of all energy are so small that when compared with nothing they are literally next-to-nothing.

It is not surprising that the difference between *something* and *nothing* as the essence of every conceivable thing is locked in an ultra-microscopic arrangement of attractive force which at even finer inspection may prove there is even a finer difference between the two..

If energy can be transposed from tangible matter into an invisible, imperceptible form and perhaps reversed it means that reality, and more

8

importantly *causation* is being effected beyond the perceptible scope of human knowledge. If something is not readily perceived and its existence is unknown due to blatant ignorance, it may be the controlling force in the relationships between all human and non-human COEs. It might be what human COEs have labeled as God. The proof for this lies in the working of situations and arrangements of tangible matter that could have only been achieved by some underlying force as yet to be defined.

All of the above mentioned forms and variations are some form of energy. There is no detectable mass for most of these as of this date so they are not considered matter. Physicists love to find ways to use these energy forms to act on other forms of matter as microwave cookers or MRIs or cell phones. They are mostly not visible but some can be detected by virtue of temperature others by instrumentation.

Although they have categorized and labeled and tested these consistent forms of energy and found some practical use at the end of the day they are still within the COE definition: a quantity and arrangement of something that can effect a change in some other quantity and arrangement of something else. Physicists are always trying to find new forms of this wave- characteristic energy and there may well exist forms that are not yet known. These have helped shape human reality like evolution and everything else. They are cited here in that context.

Solids, Liquids and Gasses

All of the above being forms of energy, conventional physics has given them different labels. These labels are for three basic groups which for no other reason make it easier to manipulate molecular formulations in theory and idea prior to an actual, practical inter-arrangement of those molecular formulations. They are only symbolic labels describing the different types of bonds within different types of matter. As the bonds vary so does the way that type of matter interacts with other types of matter including human COEs.

Generally speaking the aforementioned light, heat and various waveforms of radiation regularly act on the solids, liquids and gasses to change them in some way. Many times a liquid is changed to a gas or solid or solid to a gas. Sometimes the liquids, solids and gasses are transformed into the waveform radiation previously mentioned. An example of this is a solid like wood plus a gas like oxygen when heated or ignited will transpose into heat energy. Conversely if heat energy is applied to iron and chromium it will fuse to a new molecular bond which is stronger, harder and more dense than either element.

This back-and-forth kind of musical chairs dance happens much of the time in nature in billions of different combinations without any help from humanity. In fact most of the shuffling back and forth over the course of billions of years has occurred without man's help resulting in the tangible universe as we know it.

Today the willful control of these various states of matter is equally as important as the control of radiation waveforms. It is the planned achievement of some end thru the enacting of energy upon some other form of energy or matter. Scientists, chemists, physicists and engineers spend a good portion of their existence with that intent.

Atomic Design

All of the above mentioned energy forms interact in a dance of constant change to form energy bonds as new atoms and existing atoms create molecules and energy by-products. Everything from the silica in a grain of sand to the ink that prints dollar bills is a function of that universal interaction.

Tracing backward to the genesis of all matter is the general contention of energy as a two-sided coin consisting of a condition of presence vs. non-presence. Tracing backward again brings us back to the number-one contingency of human reality as it relates to pure physics in that everything that ever was is a COE, conglomeration of energy. Therefore all reality is a condition originating at the sub-atomic level.

If it were possible to zero-in on matter in any form by way of visual inspection, temperature monitoring, electrical charge, radioactivity or any other energy monitoring process at this level, at trillions of times normal human sensitivity, matter would not be the stable, perceptible substance we normally know. We would see pools of change happening trillions of times-per-second. In time-lapse the activity and interchange would render the material to one of constant flux not different from a waterfall.

The only thing that keeps that matter as a three dimensional construct we can count on as a rock, pebble, grain of salt or speck of dust is our relative size and our relative time-based longevity as human beings. This is why we can label something as physically real or not.

Time

Words and Definitions: TIME: Time is defined as the progression of events within existence. The occurrence of things in a series of happenings.

This definition is another chicken or egg situation. If you look up the words *occur, happening* or *existence* you will find they define each other. The defining themes of many of these quasi-theoretical, philosophical concepts always kind of dance around the point and expect the reader to say to himself "well you know what we mean."

Physicists on the other hand insist that time is a function of space and velocity and can be warped pending gravity and other influences. An equally confusing arrangement of words.

To the general public time is minutes or hours or days as determined at least by face value of the positions of the earth, moon and sun.

To the horologist time standards can vary pending which way you want to "measure" time. The atomic caesium interpretation is thought to be the most accurate (or nearly the most accurate). It's based on the predictable behavior of sub-atomic particles in caesium.

Theoretical physics has taken the concept of time one step farther and posits time is a flexible dimension which can be stretched slowed pending things like gravity, velocity or other forces.

Within the scope of conventional physics most of that stuff is either a functionally-fit definition for conventional reality or pure speculation outside of basic logic. Time is really not that complicated.

In the scope of reality as defined by pure physics the word "time" is understood as merely a symbol for the detected change in some form of energy which never stops and had no start. If that energy change is slowed or sped up by gravity, velocity or other forces of arranged energy it still cannot be considered as uniquely influenced by some time warp or super normal phenomenon. This is because the change can only be considered against a universally accepted measuring stick, whatever that might be, subjected to the same forces. In other words time only exists because we define it. To take it one step further time can only be measured, or tested or sampled because a standard of measurement has been defined. It's the same as mathematically saying 2+2=4 because that is the definition of "4" and for no other reason. Therefore if you put a mechanical clock in a refrigerator it may slow down because the molecular, atomic determined properties of the materials of which it is constructed are affected by the temperature drop. It does not mean time has slowed down.

If a caesium clock, as the standard measuring stick, shows a different rate of atomic activity in normal temperature than in a lower refrigerated temperature you can say a caesium clock in the refrigerator was slowed but time itself was not by the same reasoning. Even though time or the speed of time was established and agreed upon to be the caesium standard that standard was established for the caesium under certain conditions. Time did not slow down but the standard was altered.

To be a true controlling factor in universal time the effect would need to transcend all independent rate changes across the board. If the corrosion rate of iron was slowed down at the same rate the cellular DNA became fragmented there would need to be specific energy changes at the atomic level to effect those changes in each case that were somehow universally coordinated.

The only problem is the energy changes to cause a slowdown in the corrosion are unique to that molecular arrangement in iron and the energy changes to effect the DNA slowdown are completely different. In order for it to hold true magic, fantasy, divine intervention or a supernatural occurrence would be necessary. Later the subject of supernatural occurrences will be addressed.

The bottom line is if the agreed upon measuring stick was decided to be caesium at a specified temperature, pressure, gravitational contingent (earth's gravity) and altitude, in a radio-shielded enclosure then any deviation from those standard ambient conditions may result in a modified interpretation of time. So even if velocity did indeed change the interpretation of a unit of time, time itself was not changed because the defined standard was established at a different velocity.

To posit time or living biology can be slowed down by traveling at a high velocity is like the government just printing more money to create a higher standard of living. There may be more dollars but at the end of the day but when you add up the collective net worth nothing has changed.

As far as normal, human reality is concerned, people waste time, take their time, stretch time, do time, take time-outs and make time as well as a host of others. Time also drags and flies. It's also expanded and contracted thru the use of psychoactive drugs. The funny part about all of that is that none of it actually happens. In reality the caesium standard (or whatever sundial has been agreed upon) does not slow or drag or fly. The rhetorical, working, everyday definitions are also based on reality, although different. As a function of human perception contingent upon external factors it is itself based on energy transfer and change which creates a unique subjective reality for that individual.

To cut to the chase all of the conjecture about time is the product of delusion. Not to say those individuals need psychiatric help but their claims are purely their interpretation of reality at a specific point in energy transition relative to the *ideological concept* of time. Along with those created realities goes the terms "beginning of time and end of time." Time is a function of the perception of energy and as far we know energy has had no beginning nor has any end.

Human reality operates within a matrix of nothing more than COEs operating on the contingency of a predictable energy movement or transformation. This could be the sub-atomic activity of the caesium atom, the movement of the sun or vibrations of a quartz crystal as in a wrist watch. All movement and transfer of energy on the planet is either tracked and tagged or planned and executed in concert with those measuring sticks.

Just more quantities and arrangements of energy interacting to some logical end on the micro side but no logical end for the big picture. If you look at the theoretical aspect of time with respect to moving forward; the predicted change in energy is

based on the historical change in energy so even the future with respect to "when" is at best a gamble.

What you find-ah
What you feel now
What you know-ah
To be real

Cheryl Lynn

Chapter Three

Reality In General

The concept of reality is important when talking about the development of humanity because all culture, values, civilization, laws, behavior and most everything else man has created as a norm is painted against a canvas of reality. If someone comments about the weather; "it's really raining" or "it's real hot outside" they are conveying the messages that the falling rain or the air temperature are not subject to individual interpretation but are branded with a property of magnitude which resulted from a majority opinion. That majority vote established a truism. Conversely if it was a feeble drizzle in the wind or a temperature that was somewhat warm-to-hot a subjective opinion would probably bridge the actual conditions into a plausible label. A message of "it's sort of hot" or "it seems to be raining a little" would be the conveyance.

But because there was no doubt about what was occurring outside in the first example the message generator has given a trusted label to the conditions by calling them "real."

Reality is in effect an agreed upon plane which the majority can use as a blank-slate for information, communication, building laws, culture and values. The ultimate result is civilization. Reality however is actually *highly subjective* but the societal condition of humanity forces this into a majority- agreed upon interpretation as a matter of necessity for the benefit of the larger group.

If you ask anyone who has had experiences with marijuana or any other psychedelics they will attest to the fact that reality is how we interpret it. A crazy person is really not crazy when you consider *their* normal biochemistry is giving them a view thru twisted glass. Memory in this case might not store information as accurately as a normal person's would. Perhaps not as long or perhaps longer. Perhaps the information is "misfiled." A sunset might be remembered in a category of TV programs or as brand of liquor. It is for this reason a criminal found guilty of a crime by reason of insanity is not punished but isolated and maintained for his natural life.

No big deal you may say but the point is reality in essence is a function of subjective biological interpretation. The same reasoning even applies to lower animals. A lion's reality is different than a dog's. In the animal kingdom however there is less forgiveness for a unique interpretation of the world, which is why you don't see too many crazy dogs, cats, squirrels or lions.

The significance of reality as related to humans and their interaction with the laws of science of physics lies in that same energy construct, the COE. If we consider reality a subset of sensory interpretation and consider the building blocks of interpretation perception and the engine behind perception to be energy, we can see how energy determines reality.

The biological process of perception is very simply the changing of energy in cells within a living entity based on external energy forces. If sunlight is cast on an apple tree, light energy from the sun causes red wavelength photons to be emitted by way of glowing molecules on the surface of the apples. This is much the same as a black-light would cause minerals to glow in a museum display. The differences lie in the light energy wavelength, the difference in the molecules being illuminated and the way the energy emitted in the form of photons is being perceived by a living entity.

If we continue the energy path into what is labeled an actual perception we see it as acting on other cells (COE's) within the retina. This in turn creates a neural energy change with a domino effect up into the brain. For all intents and purposes let's say the neural energy change within the brain is maintained for X amount of time which man has established as some threshold limit. Now we can label the condition as a memory.

If the living entity closes his eyes or thinks back an hour later he can see by way of mental imagery that same apple tree in the sun. Still no big deal as basic biological processes are subjected to the physics of light. The big deal is the difference between the initial illumination on the tree as perceived and the image which is sustained in memory. Which is real? Which is reality? If you consider what actually occurred in both cases the similarities present a striking question. In the first case energy (light) acts on a COE (the tree) to produce another energy variation (light reflected) which is perceived by the a COE (the human being) thru neural energy channels. In the second case an alternate energy force (the neural memory) acts on a COE (the brain) causing a perceivable image by a COE (the human being himself). The difference in the two being a function of time, at least in one respect. The first image created by the sun is immediate and subject to change as attention to other external energy comes into play. The image in memory is "timeless" in that it is similar to a photograph. This is not true however in the strictest sense because memories are subject to degradation and change but that is irrespective to comparisons between reality and non-real objects. Also the initial image as reflected is only immediate as far as the human definition of immediate is concerned. There is a time lag for the whole thing to take place although extremely minute. So reality is at least a function of perception and time. In relation to energy, the building block of all things, reality is snap-shot of energy.

Words and Definitions: PRESSURE: Pressure is defined in conventional physics as a force in a specific direction applied to some associated surface. Force is defined as anything that can act on something else enabling it to move from its position in physical space.

Physicists go on to micro-analyze pressure and force using mathematical formulations enabling them to predict how materials will change or move under defined conditions of pressure.

In the world of pure physics and how it impacts human reality, the definition is much simpler. Essentially pressure is energy. That energy for whatever reason may interact with or against other energies or COE's thereby effecting some change in all energies involved. This condition starts as the basic attractive forces within the structural energy building blocks of matter. This result is a denser or more efficient arrangement of force sufficient in quantity to overcome any resistance by other arrangements of energy in close proximity. The net result is a form or shape of matter. The relationship of that dense arrangement of energy to a human COE determines a qualification for reality. Reality is once again defined thru human perception this time based on pressure exerted between the dense matter and the human COE's perimeter energy arrangement. More simply labeled *tactile sensation.*

Referring back to the example of the apple tree. it can now be argued that the apple tree in the first case being a visual entity as opposed to its image in memory and in the second case cited by virtue of the apple's capacity to be touched the apple is real.

The touching of the apple is in an example of a neurological process based on pressure. The tactile pressure results in a neurological change of energy illuminating the brain with a similar "image" to the visually produced one. If the apple or bird or whatever the object in question had also produced a soundwave it would further support a perceptible impression of something which is real.

So it might be argued that the greater number of *different perceptual modes* used to sample a proposed reality construct will provide more support for establishing that particular construct as a true reality. This is all well and good

17

except if you consider someone without sight or hearing. In any event reality is determined by perception and perception is a function of energy.

Because both the apple tree and the image in memory of the apple tree are comprised of energy, the only actual difference between the apple tree in the first case and the image of the apple tree in memory is in the *quantity and arrangement* of that energy. If those two factors are discounted reality does not exist or exists without distinction from non-reality within the framework of pure physics.

It might also be argued that of course the example with the tactile features or sound waves is the true reality unless of course you are crazy. But then you need to look more closely at the definition of crazy.

If the majority of people choose to vacation only under a full moon because it has a more pronounced relaxing effect and businesses are closed during a full moon, the non-conformist would be labeled as crazy for vacationing otherwise. Crazy is also a majority-defined condition. Once again the human majority makes the rules as to what is black and what is white.

Back to pressure as related to human reality in the modern world as we know it. Human beings often choose to "escape the pressures of reality" thru the use of alcohol or drugs or even a specific activity like bowling, swimming, yoga or traveling. Only figures of speech, right? In this case the "pressures" cited are not a direct physical force with tactile properties but are none-the less quantities and arrangements of energy that can be perceived more often than not thru sound and light. A voicemail, an email, the morning alarm clock, a signal to prepare the family breakfast, a request from a customer, a bill in the mail, a college exam to prepare for, etc., etc., etc.. For the most part light and audio, which is in the form of waves, are changing neural energy to effect a change in memory which further executes a change in neuro-muscular energy to effect yet another change in light, audio or physical pressure in the form of movement.

The bottom line as usual, is a suite of quantities and arrangements of energy acting on other quantities and arrangements for some necessary purpose as defined by humanity. Although interpreted on an imaginary plane those "pressures" form reality, crazy or not.

The escape from reality or "the pressures of reality" occurs by way of some kind of respite from the fatigue produced in those energy transpositions. This escape reduces the normal energy transformation processes to a point where the biological alarm warning "fatigue" can be switched off. At the same time task-specific energy reserves are being replenished. In all cases task related memory and information processing is left idle and reward centers are stimulated as replenishment is genuinely taking place. This restores *homeostatic balance*.

In situations where psychoactive drugs are used reward centers are stimulated as normal neurological energy channels are bypassed. This bypassing may have the same beneficial replenishment effect as other recreational activities because those channels can now focus on replenishment rather than operation. The negative aspect of this is that the motivation for the ingestion of nutrients may also be "switched off" leading to added biological degradation.

In the case of some other recreational activity being used as the escape, the relief also comes from cessation in information processing and memory which was task-specific thru an activity of distraction. Replenishment of energy reserves at the neural and muscular levels also takes place. This is sometimes enhanced thru physical stimulation as exercise.

As far as a basis for reality is concerned some aspect of energy is perceived and experienced on a subjective, individual basis and established thru rules of life. One of the reasons civilization was formed was to establish a measuring stick for what was real and what was not thru the eyes of humanity. Pure physics makes no distinction but acts as a strict conduit to that end.

They tell us that
We lost our tails
Evolving up
From little snails
I say it's all
Just wind in sails

Devo

Chapter Four

Evolution 101

***Words and Definitions: EVOLUTION*:** From early Latin evolve; To unroll or roll out. An unfolding of present position into a new open position.

Everyone knows what evolution is and its counterpart creation but because the book itself "evolves" primarily from the evolution perspective a brief synopsis follows:

Perhaps 15 billion years ago scientists say our universe was formed and 10 billion years later the earth. Life on earth in the form of microbes, bacteria, algae, etc. dates back about 4 billion years. Man did not appear until millions of years back and is a relative newcomer. It's ironic that the measuring stick of time to tag all of these events did not come about for 10 billion years as there was no earth to spin. For anyone to actually make note about that spin as related to earlier events took another 5 billion years.

What caused the origins of all this stuff, considered both non-living and living, can once again be attributed to COEs. COEs haphazardly came together over billions if not trillions of years tossing and interacting in infinite combinations. Eventually a detectable physical substance which we call *matter* formed what we now call the earth and planets. Just because the energy "solidified" from a human perspective it did not stop the continued intermingling and interaction of the varied conglomerations of energy. What followed was the orbits, seasons and weather as we know it.

On earth the energy toss over the next billions of years put together a puzzle by which the present arrangements were able to absorb other energies or COEs in immediate proximity. After absorption and integration of those COEs the original COE was able to use them as building blocks to create a duplicate of the itself. When the energy reactions reached this unique capability it was termed "life".

Granted there have been and are indeed today COEs which are neither living or dead or both living and dead, such as viruses, however these are still constructed of basic energy.

As the original models continued their replicating behavior they sometimes lost accuracy in the duplication process. Some of those errors in replication actually produced a more sustainable COE with respect to its proximal ambient energy matrix (environment). This process of proliferation and modification went on until the earth was populated with COEs of all different arrangements but still retained the capability of self-replication. Those particular COEs could for the most part continue the manufacture of that model much the same as auto manufacturers continue their lines of Chryslers, Toyotas or Hondas.

In any event present day man is considered one result of that process. So man has been "unrolled" in the process of evolution and we have labeled him as the final chapter of a scripted play even though nothing seems to indicate the process has stopped with his genesis. As life is sustained even longer thru medical breakthroughs more versions of man are surviving to the age of mature procreation. Some of those survivors undoubtedly may have cognitive capabilities far outside of the norm like some of the demonstrated musical, mathematical or memorization genius we seen today.

At the end of the day present man is still a COE no different than any other COE in the relation to pure physics and may just be a stepping-stone to that next level of change although imperceptible thru his thin slice of time.

The Lower Forms
As an introduction to human forms of life it is necessary to get a perspective on what are considered the lower forms of life. The definition itself gives some insight into the attitude towards a purpose for humanity as concluded by none other than man himself. If the majority of the animal kingdom is lower and man is higher he is indeed the king of the universe.

The terms lower and higher do not actually refer to a physical height above the earth as we all know, but even if it did man would still be the king because of space travel. In fact space travel itself might be an act of prideful advancement to support the same contention that man is the highest form of life. The proposed motive of curiosity may very well be a smokescreen of humility to hide that sense of pride.

Man however is generally referred to as the highest form of life in his eyes because of his *intelligence*. The only problem is the intelligence scale he is measuring all life by is his own design. If the only tool in the box is a hammer it's very easy to label everything else as some kind of nail. Without opening that can of worms it can be considered for all intents and purposes that man's intelligence is higher than a ground squirrel. Unless of course the test is finding buried acorns in

22

the dead of winter.

Starting at the bottom of the life forms scale and based on the definition of life itself there is little difference between rust growing on a piece of iron and fungus or bacteria growing on a tree limb. The chemical structures at the molecular and preceding atomic levels are different but that aside the two active situations of existence are very similar. As we proceed up the life form scale we encounter life with some kind of *physical form*.

The term physical form refers to an arrangement of perimeter energy which provides some resistance to outside energy forces capable of entering the life form itself. Further up the scale we find specialized cells consolidated together to perform some specific process necessary for the support of life of the living entity itself. Teeth, mouth, heart, muscle, bone, etc. are all specialized to do their part. Specialized or not, at the end of the day the living entity does the very same thing the fungus did: absorb specifics from it's proximal environment, get larger in size and produce a replication of a similar consolidation of cells.

As life forms evolved however the replications were somewhat imperfect. This was the result of natural differences in environmental chemistry as well as wave energy events. Combine those wild cards with DNA subject to natural degradation and it's easy to see the opportunity for diversity within any particular species. Even today mutations occur due to those determinants. Eventually life forms mutated enough to be able move greater distances, at first in water, and this was considered a step up the ladder. Once again the step "up" is defined by the guy on top. In reality it was just a mutation without value or non-value.

When fish became an air-breather and developed small legs this was considered another step upwards towards humanity very simply because humans have legs. Other features came about which were also considered important like bearing live young and an opposable thumb and these were similarly used to rate a dog or cat as more worthy of human attention than a frog or fish.

Eventually life forms acquired unique properties neurological activity which man labeled as intelligence and emotion. Up until then the lower forms based their regular actions and activities on what is termed instinct or an even more basic knee-jerk type of neural reaction which merely facilitated survival. This may have been a defense reaction, killing for food or unique biological reproduction.

When living things had neurologically mutated to the point of seeing beyond the satisfying of an immediate need in favor of mapping out a future executable plan it was considered a milestone. The building of nests and even using some tools like a rock to break open a coconut further distinguished the latest life forms from a clam or earth worm.

Some sense of communal responsibility also helped to give living things a raise in status. Caring for the young and hierarchies of decision-making and leadership are common throughout the animal kingdom. The sense of compassion which human beings find so characteristically "humane" is not completely absent in the lower forms. If an offspring is lost or taken from the mother in many lower forms a true sense of sorrow is detectable although it may be short-lived. Even ants will band together in a sense of empathy to free a fellow ant trapped under a pebble. Ants will also farm mold from seed spores for future use. They have also been known to apply crude, natural antibiotics to ailing ants in an effort to cure their ills. Conventional news stories often cite a dog or a cat who has summoned a neighbor for help when a human emergency has demanded it. Animals will also nurse a completely different species if the actual mother is absent.

In spite of all this stuff and countless other examples the lower forms just don't make the grade when it comes to measuring up against a genuine human being with all his bells and whistles.

At the end of the day however all the king's horses and all the king's men can't deny both lower forms and higher forms of life are still examples of energies in conglomeration making their way across a dense matrix of other ambient energies.

All together now, God made man!
But he used the monkey to do it
Apes in the plan
We're all here to prove it
I can walk like an ape, talk like an ape, I can do
what monkey do
God made man
But a monkey supplied the glue

Devo

Chapter Five

Human Genesis

Long about some millions or so years ago the chimps or monkeys mutated to the point of being labeled human. Granted there may have been some crude versions of the human being at first production out of the trees but they were the closest thing to modern man to ever walk the earth.

In light of the age of the universe, or even the later ages spawning life, the span of human genesis is a thin slice of time in its infancy. This lends to speculation on the future of the evolving species; if the earth lasts that long anyway.

If the concept of evolution is the accepted rule there is no reason to posit anything other than a continuation of rogue mutation resulting in a being so different it warrants a different species label. There would be however one caveat based on man's presently evolved intellect and that is one of self-determined value. Because man has the capability of conscious awareness he assumes mutation in any direction would be a correct, significant advancement and will rewrite science to fit that definition.

If certain mutations like a propensity for a certain disease are corrected genetically the primary direction for offspring will be determined as positive but secondary effects, generations away may be hindered. No big deal except this is uncharted ground. By the nature of evolution itself it's always a dice-toss. If man takes steps to shape his species or even his nutrition by *GMOs* in an effort to create a reality he has determined as advantageous and those steps inadvertently retard the future, the lost ground may never be regained. Who is to say three eyes are not better than two? The human population on earth would probably be double what it is now if a third eye had mutated in the back of man's head a million years ago. Unfortunately though this has not proven so in nature as some spiders do have six or more eyes.

Words and Definitions: HUMAN BEING: A man, woman, or child of the species *Homo sapiens*, distinguished from other animals by superior mental development, power of articulate speech, and upright stance.

Starting with the easy ones, upright stance is defined as the stance humans have. Giraffes are certainly upright but that doesn't count. Oh what about upright on two legs as a qualifier? No dice! Birds even sleep upright. How about articulate speech? Until somebody can translate a dog's bark or whale sounds or bird chirps it's inconclusive at best. The species label is another chicken or the egg definition. It really does not give any insight into the subject. The icing on the cake is the superior mental development. If humans are capable of anything it's a superior intellect.

Man's extraordinary mental abilities stem primarily from his memory and imagination. Lower forms of life have both but they are not exaggerated to the point of mapping out the distant future as they are with humans. There is also a somewhat impractical behavior which humans effect thru their intellect and that is artistic creation.

A spider will spin a web or bird create a nest but it is for a normal, practical purpose and not entertainment. Instinctual or not man will create something that starts in his imagination for no other reason than it gives him or someone else pleasure to do so. As art imitates life man tries to recreate symbols or images of reality as he sees it.

Man will also use his intellect for a perceived necessity in the future based on his memory of the past. He will carry that perceived necessity into a physical action to create "closure" in his memory, thereby neutralizing any accumulated psychological tension. That tension might be loosely defined as *conscience*, another feature most of the lower forms lack. To simply illustrate, if swimming is the afternoon's intention, packing a bathing suit is a necessity. That necessity will set up a condition of tension until it is packed. Lower forms generally act on the necessity of the moment rather than future necessities. Up-the-ladder forms of life having concern for the future is more prevalent than a squirrel burying nuts for the winter.

Man also has the ability to create a social network of behaviors and activities based upon an agreed upon set of rules which he calls *laws.* It's true there are laws of the jungle but these are for the most part directly determined by the forces of nature and instinct rather than conscious decision.

The label of "law" in this case is once again a man-made idea but in reality it's very simply the same old push-and-pull of energy to one end or the other. The lower forms either conform or suffer or die or both. Man has the ability to conceptualize, to set up hypothetical conditions based on memory and predict a possible outcome. This is by far his greatest advancement over the lower forms of life.

This capability is often **overshadowed** while emotional overtones are heralded as the sacred difference which separates us from them. Compassion, charity for the poor, care for the sick and disabled, extending life of the aged as well as justice for the innocent have dominated man's claim to fame. The reasons behind his newly acquired sense of superiority is to be discussed later. At this point it is only necessary to define the demarcation. That demarcation will determine reality as dictated by pure physics and reality as defined by humanity.

There is a commonality with lower forms that cannot be escaped by virtue of outgrowth. In order to get where man is today on the evolutionary scale he had to retain some of the features which his predecessors possessed and lower forms still possess. It's for this reason human beings are still part of the animal kingdom and will probably stay that way at least for a while. A being which is both biological and of a higher cognitive ability as defined by man himself are two halves necessary to make man human.

The Biological Half

There is no actual biological half when it comes to human beings, it's all biological. In fact *biological* is an all- encompassing term for bio-chemical. To take it one step further bio-chemical analysis is structured around bio-molecular. Then comes a generic flavor of atomic, sub-atomic and some version of energy. The sub-title *Biological Half* is used because we as humans interpret the higher-order thinking processes as non-biological but in fact they are biological. If you substitute the term neurological for biological in the above scenario the end result is the same, as some kind of energy determined process. For the sake of clarity I will refer to two aspects of living things and not one.

In any event the biological half with respect to pure physics is a no-brainer. In fact there is little difference between what goes on in a fish or frog and what goes on in a human being.

All things we label as living are in essence some kind of biological "tube or vessel" which is in essence a COE comprised of a multitude of other COE building blocks.

Words and Definitions: HOMEOSTASIS: is the state of steady internal conditions maintained by all living things. This dynamic state of equilibrium is the condition of optimal functioning for the organism and includes many variables, such as body temperature and fluid balance being kept within certain pre-set limits as a *homeostatic range.*

The major (multi-assembled) COE cited above as biological is a homeostatic system of both energy in the form of matter labeled protoplasm and other dynamic quantities of energy in the form of chemical, thermal or electrical. The definition of homeostasis references "pre-set limits" and it's interesting to wonder as to who set them? The big player here is evolutionary development within the constraints of pure physics.

An environment void of the necessary energy quantities and arrangements to maintain that dynamic equilibrium will be either exited or *added to* as a function of the dissemination of the living entity after death. The pre-set limits clause is another chicken or the egg issue in which the existence of the limiting factors and the immediate environment determine each other. Since neither came first or last the matrix of time cannot be applied in the strictest sense. At best a weaving of adaptability vs. non-adaptability took place.

In any event the COE, a structural "tube or vessel" we labeled as living must function at the biological level to qualify for that label. The major biological activities of a living being consist of extracting nutrients from the immediate environment, processing those nutrients and excreting some of them back into the environment. The nutrients that are not excreted directly back into the environment are used to either add matter to the living protoplasm (tube/vessel) or substitute protoplasmic matter for that which has been excreted indirectly or over extended periods of time as replacement. This is why the living entity is in effect a tube which retains some energy but passes other energy thru in the course of its life.

Eventually all nutrients extracted from the environment in their entirety are returned to the environment by virtue of the degradation of every building block which makes up the tube itself. This is labeled "death." It should be noted that the extraction of nutrients from the environment is the number one priority for all living things.

As a matter of the purely physical arrangements and quantities of energy within the organism, whatever it happens to be, the organism requires replenishment and maintenance of those arrangements. To reach this end it will respond with attraction towards a nutrient source. In most cases this attraction is initiated thru a fine-particle "pilot" such as free-floating molecules from the nutrient being absorbed and *sensed* by the living COE in need. Sensing is in effect testing the molecular sample thru specialized organs which for the most part were hard-wire programmed by way of genetic design. This instinct like reaction may also be accentuated thru learning or past experience with similar molecular samplings by virtue of comparison to a memory trace. After sampling and accepting the sample as a correct, needed nutrient, acceleration of the movement of the COE thru other energy fields is enacted. This mobility towards the nutrient source takes place until the nutrient source is reached. Subsequent absorption and extraction of needed energy building blocks results in a renewal of homeostasis. This happens in any living thing from a tree root burrowing out for water to a human duck hunter.

The question of consciousness might be relevant in the above example and that will be addressed later, but the short answer is nobody knows what the tree is thinking.

In the final analysis as far as pure physics is concerned a COE has intertwined with the matrix of energy it was created in, absorbed and processed some of that energy to produce other COEs and non-conglomerated energy forms to maintain the condition of life. Inevitably later they were transferred back to that same energy matrix in death.

The only other major activity the biological half pursues is the replication of the major COE assembly and this is known as *reproduction.* Reproduction involves the same environmental energy extraction as in the nutrient process except certain quantums of energy are channeled for building blocks of a replica of the major COE. This is done within the defined energy border of that major COE. When the replica has been allotted enough energy constructs to enable it to physically extract nutrients from the environment it is moved thru the major COE, thru its defined energy borders into the environmental energy matrix. In the case of eggs the COE's energy perimeter is crossed but there is still a time lag before the replication actually starts extracting from the ambient energy matrix (hatching). In any case it begins its own "life" and goes on to reproduce again at a pre-determined point in time. This activity happens in all things man has defined as living from bread mold to elephants. The mechanical process by which this occurs may be perceivably different but it's all essentially the same.

In humans both sexes are required at least in some form to set the stage for the same basic process.

The big contingency in everything from cellular biology to animal husbandry to human gestation is the time lag between birth and replication. Realistically this is the *replication-interval.* For whatever reason the human intellect is obsessed with consciously tagging the rate of replication against the normal change of universal energy (time).

30

To consider every situation in summation, it can be concluded that when any COE from a microorganism to a human being has extracted and absorbed a "pre-set limit" of nutrients from the environment and has added a "pre-set limit" of new protoplasmic matter to the major COE, replication is enabled and will take place. As stated earlier changes will be effected in the building blocks created by those nutrients and protoplasmic matter within the defined energy border of that organism will take form in the necessary quantities and arrangements of energy to build a replica. In the case of male and female contributors as in humans the same reasoning holds true for each on an individual basis. At that time a joint effort to bring the necessary building blocks together is achieved and replication follows.

The dynamics behind the culmination of the replication stage are nearly identical to the nutrient extraction process for the sustaining of life. At some point a homeostatic alarm becomes perceptible within the organism signaling the need for reproduction. The longer the time without a mating partner the louder the homeostatic alarm sounds. At this point sensory systems are activated beyond normal to increase detection rate similarly to when nutrients are needed. Sight, sound, smell and touch are accentuated. These are also influenced by ambient energy influences as temperature, light, wind, etc..

In the case of nutrient acquisition smell in ambient air is important in that smell and actual ingestion are linked in sequence. In light of pure physics this is a mechanical process.

A molecule of the desired nutrient in its own arrangement of energy is scattered thru the ambience of the environmental energy matrix. Being inhaled it excites an energy pattern within the organism which not only further sets off the homeostatic alarm but also enhances all other senses still further to detect direction and ultimately location within the environmental energy matrix. Neuromuscular activity moves the organism closer to the source.

In the case of mating partners the scenario is basically the same except visual signals probably count for more initial excitement and motivated mobility. In this case light energy in the form of photons is radiated from the candidate and captured by the opposite sex as a neural image. The neural image then does all the things the scent molecule did within the organism. How much of this is actually hardwired and how much is developed thru higher order learning is speculative at best but evidence in lower forms lends most of this process to be "hard wired" which can be labeled as instinct. In fact it's just a systematized arrangement of energy constructs that got there by way of evolution and chance.

The questions of choices and decisions, preferences as well as deliberate over-activity and under-activity will be addressed in *the cognitive half* with respect to human behavior and that's where it gets complicated. Lower forms simply satisfy a homeostatic need.

The human motivation for reproduction has been adulterated by man-made cultural norms and laws and much of the time it is reduced to a crap shoot or analytical puzzle. After you clear away all the rhetoric however it still comes down to the pure physics transposition of energy.

The Cognitive Half

Before proceeding it might be a good idea at this point to present a time-line summary of what has transpired in the path to an understanding of basic human reality from the perspective of pure physics:

1) In the beginning (which really can't be established but we need to start somewhere) there was pure energy in a multitude of forms, perhaps trillions of years back.

2) The matrix of different energies intertwined and reacted to form matter. Subsequently the planets including earth were formed billions of years back. At the sub-atomic level energy was still energy and subject to constant change and transformation without physical or palpable properties as we know them. The crossing point was the conglomeration (COE) in sufficient quantity and arrangement to effect a physical mass that could be measurable with respect to unaided human perception be it the earth, gold or anything else.

3) About four billion years ago various factions of energy on earth still intertwining and the inter-reacting created an arrangement of matter which had the unique properties of extracting nutrients from the environment and replicating itself as a major, definable COE. Matter with these properties was labeled as "living."

4) Over the next three billion years the living variety of matter had replicated into thousands of humanly perceptibly different COEs. Some of those COEs had unique features which inspired a characterization of the different groups based on the defining border (physical traits) as well as the activities, methods of replication and the specific environment they were able to extract nutrients from. These were the lower forms.

5) Within the last billion years the COEs had mutated during replication to the point of being labeled human. The human being is not different from the lower forms with respect to a certain group of traits considered biological but they are with respect to a set of traits acquired by mutation that are

considered *cognitive*. Granted lower forms do retain some form of cognitive functioning at varying degrees but what the human being has determined the qualifiers for those traits to be of a higher order he has ascribed to a certain set of defined abilities curiously characteristic of humans. It may be a battle of words but is the accepted standard for classification.

Words and Definitions: COGNITION: the mental action or process of acquiring knowledge and understanding through thought, experience, and the senses. From the Latin "to get to know."

Most lower forms of living things have the capacity to think and in fact some are extremely clever. Bees can build hives, beavers dams and birds' nests. Ants can organize labor and even medical care in ant colonies. Dogs, cats and horses will learn and entertain in exchange of a special edible treat. Chimps use tools and some species of birds will actually drop pebbles into an unreachable level of water to get it to rise within reach. Lions, and other predators use their intelligence to trap a meal.

Nearly one-hundred percent of the motivation behind these behaviors is to maintain some form of homeostasis which will either ensure the longevity of the organism or its replication. If the behavior is group-advantageous the homeostasis has at its core a hardwired genetic program called *instinct*. Instinct is basic energy processing within the laws of physics but may be influenced by subjective choice. In any event all of these behaviors demonstrate some form of cognitive ability. The difference between humans and the lower forms with respect to cognitive ability is that in humans cognitive ability is on steroids. No pun intended but in fact it may be due to some bio-chemical or hormonal influence as well.

If a horse, dog, cat or chimp were subject to adverse conditions pushing the homeostatic balance out of limits an immediate response would result to correct the situation immediately. In the case of human beings the response would be buffered or mediated thru priority filters and options. Those would be mentally projected thru hypothetical mediation against a matrix of time. After the options are balanced and analyzed, action will take place based on the *best possible choice*. In essence this is thought, planning and execution.

Lower forms do exercise somewhat rudimentary similarities but they are limited in time (immediacy or within a defined cycle like a squirrel burying nuts) or repetitive. This is basically impulse determined as a function of instinct.

Human cognition has the unique power of anticipating the future and acting in accordance to maintain some kind of homeostatic balance for a condition that does not yet exist (imaginary plane). This presumably puts him at an advantage over the lower forms because he can somewhat create his future reality as he goes along. This is in the ideal sense because observation of man in the general sense has historically pointed to a plethora of errors in not only planning and executing but reasoning as well.

Man also has a superior memory function in that he can store all kinds of information, facts and measured conditions once again in anticipation of some need in the future.

Imagination is the engine behind any kind of planning and may have either a factual, practical purpose or simply speculation bordering on fantasy. Taken to an extreme, dreams are a form of involuntary imagination. The funny part is animals also dream. A cat might dream he is chasing a mouse or a dog chasing a cat but that is as far as it goes. Humans appear to have some conscious control over this mental process (imagination) and can generally summon it at will. This also establishes a creative aspect of human behavior in the form of art. Lower forms do not create without practical purpose. Humans use art mostly for the purpose of invoking a visual effect. Humans use any combination of practicality and aesthetics to create things. A good deal of these creations are palpable matter but a good deal are not. Some of these are intangible though communicable by way of ideas, laws, theories, designs or even complete fantasies.

Man has a large mental "drawer" which he can file these things away in by means of an electrically charged memory trace. Man also has the ability to analyze, sort and assemble both material items and intangible items which is something the lower forms are incapable of with the exception of nest building or gathering food.

Under the intangible heading comes other cognitive -dependent processes like judgment and justice, the ability to take dissimilar facts and draw some kind of conclusion or direction from them.

One of the most differentiating cognitive factors that defines man is his speculation, belief or denial of the existence of God or some form of supernatural control which is presented as intangible. A combination of imagination, faith, physical evidence, conveyed history and knowledge as well as personal experience through active situations and emotional (spiritual) feelings are the driving forces that make God and religion an enduring part of the history of humanity.

Across the board the majority of organized schools of worship entail maintaining an account of human behavior over the course of a lifetime to determine a course after death. This is unique to human cognition.

The lack of logical accountability for real, perceptible events both enhancing life and upsetting to life have always been attributable to some "higher power" for the human majority. By the same token a smaller portion of the human population does not recognize any events, explainable or not by a divine power. It's for this reason religion can only be considered speculative even though some humans have chosen to die before denying their allegiance and beliefs based on anything from pure faith to actual experience with divine intervention.

In strict contrast the lower forms have not presented any evidence of an awareness at all that God may exist or not.

In conclusion it is established that accelerated, more elaborate cognition is evident in human beings as opposed to the lower forms of life. This is interpreted as a step up or higher in the evolutionary chain based on the logical premises of human-created philosophies. If you could ask a dog who was higher on the chain he might respond with dogs because man does not have the highly developed sense of smell and neural processor that goes with it. Man needs to analyze chemical compounds thru elaborate electronic and bio-chemical apparatus which took him thousands of years to perfect in order to detect a disease such as cancer. A dog can unlock those biochemical tracers in seconds using his innate reasoning.

In any event for the sake of sanity the human intellect is considered higher. Also in conclusion it can be established that once again, even something as intangible and theoretical as human superiority in cognition, real or imagined is simply a function of energy transfer and transformation. From light and visual perception to sound in words symbolic, tangible matter is transformed into a neuro-chemical energy trace which is submitted for analysis, memory, imaginary reconstruction or recombination and used to execute an auxiliary action towards some end. The lower forms just can't do it.

Has the dawn ever seen your eyes?

Have the days made you so unwise?

Realize, you are

Emerson, Lake and Palmer

Chapter Six

Human Reality Within The Spectrum Of Pure Physics : Overview

Reality can only be understood as a subjective experience. What is real for a dog or cat or mouse or amoeba is contingent upon their place in the environmental matrix of energy and their interpretation of that energy. We can all agree to a measured temperature or atmospheric pressure or incidence of light but how it relates to one or another living entity is entirely subjective. This is especially true considering the measuring sticks for much of what is determined "the real world" are the products of intellect of that very same living entity.

Shortly after the dawn of the human race man had recognized the need to create a unified perspective of the real world so as he organized into tribes and subsequently what has been labeled civilization he began to define reality.

Words and Definitions: CIVILIZATION: the stage of human social and cultural development and organization that is considered most advanced.

So as the organization of people became the most advanced with respect to human capabilities it can be subsumed under the umbrella of whatever reality was understood at that time that man acted not only subjectively but in group accordance to create this sub-structure with some intent of benefit or purpose.

For all those considering the real world as understood at that time, the human idea of civilization was both voluntarily and by-force secured based on majority intent. This collective reasoning had formed the foundation for laws, customs, culture and values.

In light of pure physics this translates into human COEs coming into existence within a matrix of ambient energy and trillions of other COEs. The job of humanity was the re-arranging of not only tangible COEs but imagined constructs like education, justice and religion. Those imagined constructs in the form of a symbolic memory trace were established thru sound waves as speech, later as writing and still later as audio and video recordings in matter.

At this point in the process tangible COE's are transmuted into somewhat tangible, perceptible symbols as words. Those symbols by virtue of light incidence or sound energy are further transmuted into a bio-electrical images, symbols and recreations in human memory.

After a consciously keyed awareness of the agreed upon universal energy transformation (time) those memories are effected in either sound energy (speech) or thru a semi-tangible medium like writing. This point signifies another feature of human cognition; analysis and synthesis.

The symbols brought out of memory and presented within a flexible energy matrix of light and sound can now be rearranged in replication of another memory trace pulled from a different drawer labeled *imagination*. Those symbols are then used in a hypothetical consideration which after majority human COE agreement can be next processed for practical application by other instrumental human COEs.

Practical application entails matching the proposed structural elements of the synthesized idea with an imagined or speculative memory trace of those particular elements as they "tangibly" exist. For example if the synthesized idea is a theatrical play, a catalogue of costumes as well as actors are needed. Those elements also need to be accessible in the real world and coordinated by memory traces of how to attract them.

Human reality much of the time is created by going from this process of idea synthesis to speculation to feature-accessibility to execution. Why this scenario takes place is anybody's guess at this point. The lower forms however would probably say it's of no worldly good. It is however a key defining capability unique to humans.

As the human species advanced past very simple hunting and gathering there must have been something prodding them towards other interests. Was the cognitive ability simply subject to hormonal or neural mutation to accelerate it? Even if that's the case reality as far as pure physics is concerned is not changed. The interesting more complicated part is how humanity interacts with the physical world to create its own interpretation.

Every blade is sharp; the arrows fly
Where the victims of your armies lie,
Where the blades of brass and arrows reign
Then there will be no sorrow,
Be no pain.

Emerson, Lake and Palmer

Chapter Seven

Human Reality As A Basis For Survival

The mechanism for human survival is generally the same as the lower forms but also on steroids. As with all living things the arrangement of COE's within the human being, mainly due to evolutionary factors, facilitates survival as the number one priority. By survival life is existent and thru life survival is manifest. It's mainly a matter of semantics but it helps you to keep in conscious awareness man created words and words are nothing more than bits of energy as symbols for still other energy.

In unadulterated nature the important aspect of survival is nested in homeostatic balance. If this is achieved survival is guaranteed. That of course is exclusive of the effects of drugs, some of which may mimic a state of homeostatic wellbeing when in fact they are doing the opposite by reducing the life span. This may be considered a nature-adulterated condition except for the fact that some living creatures do become taken in by toxins which have the same effect.

When in serial-time the subject of survival is addressed is another question. Human survival is both an immediate issue, like the lower forms, and a speculative issue for the security of the future. The latter taking place on an imaginary plane.

The basic engines running the survival game are they themselves nested in pure physics. In the case of maintaining a functional physiology, energy is called to action based on human memory or symbolic entities like words and language mediated thru light and sound. The most prevalent required process here is to bring nutrients into proximity for absorption.

It could be next week's grocery list or a purchase requisition for a ton of seeds for this spring's crop. The lower forms of life generally do not do this however there are exceptions as a spider might spin a web in hopes of trapping a meal and ants have been known to farm mold. As far as humans are concerned preparation for the future need of nutrients is high on the priority list and is understood as a general characteristic of the species since it evolved.

Essentially the mechanics of the energy transfer process is consistent across the board. By way of a proposed, imaginary need that is not actually experienced in the present, a course of action is planned out. That imaginary need is first generated by visually witnessing another human COE suffer the effects of a lack of nutrition and subsequently experience a homeostatic imbalance or ultimately cease to maintain the metabolic functions necessary for life. This is described in the present but in reality it is an established, agreed upon fact which only functions as a symbol in memory. Obviously humans do not go around forcibly maintaining a starving population to remind them to grow crops. It might however be speculated that starvation in the world does presently take place because of a necessity to "remind" the majority of the human population starvation is a real threat. It has not been proven, but anything is possible.

An empathetic element of concern which is characteristically human is also necessary to carry the basis for survival issue into reality. Just for the record empathy does occur in the lower forms but is not as pronounced and consistent as with humans. The mechanism behind empathy lies in other mutated cognitive functions like extended memory, emotional provocation and probably *confusion.*

As one human has witnessed the lack of homeostatic balance in another individual, by way of starvation the visual and auditory image of that poor soul is compared with their own self-vested image in memory. This confused similarity produces an emotional response, partly instinctual and partly learned, which produces a mild state of homeostatic imbalance in the observer. This response-set is labeled as *compassion* or *empathy.* It is the reason why one would wince at the misfortune of another. A hypothetical arrangement is set-up on an imaginary plane where the observer and the observed might exchange places in real life. This prompts action based on a quasi-delusion.

This effected homeostatic imbalance experienced while contemplating starvation is relinquished once the "plan to ensure an adequate nutrient supply" is executed.

As far as what has happened at the energy level it's basically the same link-block procession of energy. A COE (the human being) is undergoing an energy interaction in the form of a memory trace coupled to a homeostatic imbalance which is also energy driven. This effects a neurological action to articulate in some way *a method* to move nutrient COEs thru the space matrix of ambient energy to a point where proximal absorption is possible. This happens just the same from

42

memory traces established in the immediate real world or on that imaginary plane just discussed.

Following the process further, the nutrient COEs were initially the product of light energy acting directly on other COEs in the environment (plants) or indirectly as a living entity that absorbed another organic COE (livestock). In the latter case the living entity was interrupted in metabolic processes to the point of becoming non-living and probably subjected to a heat-withdrawal beyond ambient temperature process (refrigeration) to reduce normal COE dissemination into the environment.

The process of refrigeration is interesting in that lowering temperature generally slows energy transformation at the molecular level making re-absorption in the environment slower but actually speeds electrical and possibly other wave-form energy in non-organic media. It's further interesting to note that prior to absorption by a human COE the nutrient package is often subjected to above ambient heat (cooking). This introduction of thermal energy will rearrange molecular energy within the nutrient so it is more speedily and more efficiently absorbed by the human COE. The lower forms do not do this.

In addition to regular nutrient-based energy replenishment survival for all living creatures including humans is also contingent upon defense. Just as the human COE seeks to absorb nutrient COE's from the environment, other predatory COEs seek to absorb the human COE as a nutrient.

As a corollary to this reason for defensive action some other environmental extraneous forces will also act on the human COE in a detrimental fashion. This happens not in an attempt to consciously or instinctively replenish an assailant's nutrients but will none-the-less cause the same end as diminished longevity at worst or a homeostatic imbalance at best. For example a flood or extreme mineral toxins like lead or wave-form energy as radiation are all examples of dangerous situations which in sufficient amounts will inhibit metabolic activity. In either case of predatory or environmental energy forces, the opposition will be met with some kind of defense only if the COE central to the danger is in a condition of awareness.

Words and Definitions: AWARE: having knowledge or perception of a situation or fact.

Knowledge takes the form of an active symbol or image within a memory trace that got there either thru intentional exposure (motivated learning) or by previous experience with opposing energy forces. In either case direct proximal light or sound energy does not produce an immediate defensive response as does in real-time perception of a threat. In real-time perception the threat produces some type of light, sound, chemical or tactile energy as a signal for an instinctive defensive response. In the case of a learned response, knowledge or awareness is most often the driving force causing a defensive reaction.

Why a human COE would discern an alarm signal from normal ambient, non-offensive energy is also a case of pre-learning but may also be the result of instinct in which a genetically hard-wired energy sequence creates homeostatic imbalance. To restore the balance the human COE will either increase distance between the threat and itself by mobility in the energy matrix or attack the threat. The attack response is almost exclusively the tool of choice if the situation involves a predator. The attack involves the physically crossing of tangible energy borders in an effort to either upset the homeostatic balance of the predator enough to create a mobile response in him or to render the predator harmless. Harmlessness may be characterized by a pronounced lack of energy, a neuromuscular advance in physical space distance(mobility), a paralytic state in which no energy movement is detected or a complete absence of metabolic processes (death).

Whether the opposing force is a toxin, bacteria, a soldier's bullet, a charging bull or extreme cold, the energy link is the same. In pure physics a COE is continually subjected to other COEs both tangible and intangible. Some will have no effect on the longevity of that COE, some will enhance longevity and third group will shorten longevity. Human COEs across the board are hard-wired to take some kind of action against the third group which is basic instinct. The enhancement of that instinct thru energy arrangements purposefully directed for survival is a dominant human activity acquired thru learning.

Everything from fire alarm systems to CO2 detectors to car seat-belts to the Department of Defense are all human created constructs of energy designed to maintain life. The whole sector of medical care is a massive construct of energy building blocks, COEs and memory traces in action every minute of every day targeting the defense against death of the human COE.

Us and them
And after all we're only ordinary men
Me And you
God only knows
It's not what we would choose to do

Pink Floyd

Chapter Eight

Family and Extended Groups

Humans are social entities. They do things in groups.

Words and Definitions: FAMILY: the basic unit in society traditionally consisting of two parents rearing their children *also*: any of various social units differing from but regarded as equivalent to the traditional family.

Nothing real fancy here as even many of the lower forms maintain some type of family structure. However what's actually happening at the level of pure physics might be slightly more involved especially in light of the vast variations in relationships constituting family situations today. Current attitudes and interactions of both family members and non-members are similarly varied and may range from extreme love and endearment to extreme hatred and even death.

Essentially one COE is *an individual* human entity. If that person is also a member of a family they are not only an individual but *a contributor* to a larger arrangement of energy which in many cases might function as an individual rather than a group comprised of multiple human COEs. This dual identity is sometimes beneficial to all parties, sometimes transparent with no effect at all and sometimes toxic or detrimental. If we consider the goals of longevity and homeostatic balance as the universal purpose of all existence for humanity, the family is an intentioned tool geared towards those ends.

Origins of family structure is another argument between instinct and learning but is most probably a combination of both. Each aspect compliments the other with the goal of satisfaction of some biological need and the projected needs of the future manifested on an imaginary plane. It should be noted that many human COEs are not family members or have had limited family membership and this in itself has not impacted either longevity nor homeostatic balance. In general terms however family structure is regarded as a positive thing which enhances longevity and homeostatic balance.

A good place to start with reference to families is with birth and childhood. A birth brings forth a new, distinct COE which is destined for an accumulation of internal matter (growth), possible replication (mating) and ultimately re-assimilation into the environmental energy matrix (death). If the new COE is manifested thru an existing family structure there is a greater probability of growth, extended longevity and the opportunity for replication. These are the natural hard-wired energy goals. Another hard-wired goal which has less priority but is none-the-less a driving force is the on- going quest for homeostatic balance. This need is also addressed in a *normally functioning* family.

A normally functioning family is contrasted with what is termed a *dysfunctional family*. A dysfunctional family does maintain the same proximal, light incidence and sound energy links between members as a functional family but it does not fully satisfy the enabling of the aforementioned goals of all members. In fact it may actually work against those goals. It should also be noted that the majority of family groups including dysfunctional ones are defined by law and labeled as such by universal human agreement.

The mechanics of how the family structure facilitates the prioritized goals of human life can be summed up in the few key provisions. First and foremost the family structure provides protection. This is especially important in the cases of new members, young members, old members and weak or sick members. It is an unwritten agreement to cover each other should the need to fend off a physical attack from a rogue COE of the human or the lower form variety.

Before a predator is able to inhibit metabolic energy transfer and upset homeostatic balance, the emission of light incidence, sound waves and other cues by the member under threat will produce a homeostatic alarm signal in ancillary family members. This will initiate an action to inhibit an escalation of the threat from the attacking COE. This is accomplished thru the usual links between the alarm response and effected neuromuscular action either in direct physical contact with the predator or by way of an intermediate COE (matter which will penetrate the attacker's outer-most energy perimeter). In this way the group of human COEs is acting as a composite, unified, COE who is acting to eliminate the threat.

At the level of pure physics the coming together of ancillary members is essentially a pooling of smaller quantities of energy to create a quantity of energy large enough and in proper arrangement to overcome any resistant energies consolidating the power of the attacker or threat.

Other aspects of family dynamics entail the providing of nutrients (food), providing of a physical environment conducive to maintaining metabolic balance (shelter) and providing a platform for "group-think." Group-think is basically the allocation of problem-solving memory among different family members to achieve that goal of solving a problem for any one particular family member or number of family members.

For example the problem might be how to secure nutrients at a remote location. This is effected by synchronizing the imagined position of relevant COEs in the energy matrix as coordinated in a normal, predictable, universal energy change (time). The problem might entail the transport of one COE member to another position within the ambient environmental energy matrix and synchronizing that arrival with another unrelated COE to be in that same proximal location. The problem might be one of instilling a particular memory trace in the neural circuitry (teaching) of a COE member in order to enable that member to effect some action at an imagined point in time or under imagined circumstances warranting that action. The idea being that the action effected by the new memory trace will either enhance homeostatic balance or extend longevity for that member or other members.

The advantage in group think as opposed to individual solution seeking is that with more human COEs the amount of memory traces focused on the issue is multiplied. This is due to a multiplied number of inputs over past experience with some of those experiences leaving a lasting memory (experience). Also when the different memories eliciting different perspectives are brought forth thru sound energy or symbolically thru writing they can be arranged in a multitude of combinations on an imaginary plane. The best hypothetically, imagined outcome of those various combinations presents itself as the solution or best action to take within palpable reality. Hopefully the problem will be successfully eliminated.

Family members might also share resources like *money* which will be discussed later.

Extended groups like work groups, sports teams, military brigades, educational institutions, clubs and administrative organizations are similar to families in that the group itself might function as a unified entity and may draw on scattered energy resources both in tangible matter or memory traces (ideas) to achieve some beneficial end for the group or an individual member.

In summary human COEs are diversified as sub-entities of larger groups which when required pool energy to create a new, larger arrangement and quantity of energy. This is true whether officially constructed, labeled and recognized by human civilization or consolidated out of basic need by chance occurrence in the ambient energy matrix by individual human COEs recognizing a need.

Chicago Green, talkin' 'bout Red Lebanese
A dirty room and a silver coke spoon
Give me my release
Black Nepalese, it got you weak in your knees
Some seeds and dust that you got buzzed on
You know it's hard to believe
30 days in the hole
30 days in the hole
30 days in the hole

Humble Pie

Chapter Nine

Laws

There are laws directly related to physics, chemistry and nature but those are not man-made. They are man-observed and classified. This is because they are generally recognized for importance within an "as needed basis." For any particular reason or need like calculating a volume of gas expansion in a cylinder they are at man's finger tips. The conscious awareness of these laws extends their use more as tools than constructs of reality. To include those laws in the defining of reality is once again like analyzing the differences in a grain of sand when comparing concrete constructed sky scrapers.

Laws which are influencing the reality of everyday life are generally considered a fabrication of humanity even though a commonly held belief is that many were originally imparted on humanity by divine intervention.

Words and Definitions: LAW: the system of rules which a particular country or community recognizes as regulating the actions of its members and which it may enforce by the imposition of penalties. From old Norse '*Lag*' something laid down or fixed as in position, unmovable.

Laws are mediated thru words. So before understanding laws you need to have clear understanding of words.

Words and Definitions: WORD: a single distinct meaningful element of speech or writing, used with others (or sometimes alone) to form a sentence and typically shown with a space on either side when written or printed.

That's the conventional definition. In light of pure physics a word is a sound wave within the ambient energy matrix effected by a memory trace within a human COE which has acted on a neuromuscular arrangement of energy. It might also be that same memory trace acting on a different neuromuscular arrangement of energy to effect a contrast in matter which will reflect or effervesce light energy as a symbol (printed words) in a human perceptible manner.

To complicate things further, that initial memory trace is actually an imaginary symbol of palpable matter in some form or matter that exists only as a condition of energy on an imaginary plane. For example the words Rocky Mountains are symbolic of a real mountain range in the mid-west. The words "mountain of work" are symbolic of an imaginary mountain made up of the intangible "work." A condition of heat or cold uses words to symbolize a real energy state.

So in essence words are symbols originating in human memory for the sole purpose of transferring that memory to another human COE or manipulating it with other memory traces. This is not completely restricted to human COEs because some lower forms do create memory traces and respond to words. Therefore even a dog has memory in the form of symbolic words.

The big deal about words and their significance with respect to law is that law is not only conveyed and implemented symbolically by words but also influenced in a *qualitative direction* by words. The difference between a successful lawyer and unsuccessful lawyer is often irrespective of the side of the argument but vested in the choice of words presenting the argument.

The important thing in pure physics is the concept of energy transference from one human COE to another by an imaginary symbol called a word. This is also important with reference to law because laws are created on an imaginary plane which presupposes an arrangement of energy and its impact on other arrangements of energy.

Laws are in essence an imaginary guidance system for human behavior for the benefit of individuals within a group of people. Ideally laws function to protect the human COE and organize his behavior for maximum efficiency. The objective of the efficiency falls back on those two most important biological motivations; extending life and maintaining homeostatic balance.

In order for man-made laws to work an *agreement* must be reached within an extended group of human COEs. An agreement is another imaginary construct of memory in the form of symbols, words and images. It is a construct in which any number of human COEs believe it to be consistently representative of a unified, identical, intangible arrangement of energy or tangible COE as matter.

For example if a group is in agreement that the grass is green, the color imagined in each individual's memory is reasonably more identical than different. The way this achieved and/or validated is by comparison in other unrelated examples each of which maintains at least that one feature of nothing other in color than "being green." From the perspective of pure physics light energy causes a specific wavelength to be reflected from assorted COEs in the form of tangible matter. That specific wavelength which is effervesced the same way across the variety of samples is labeled as a symbol for storage in human memory and also so that it might be communicated. The word-symbol "green" is paired with an actual retinal excitation caused by the wavelength. The wavelength associated with green could just as well have been labeled pink or snow or bird. It just so happens in this case it was creatively established because of similarity to other words in old English and German for *grass* or *grow*. In any event the point is the symbolic label for the wavelength is agreed upon and tested by way of multiple examples. The same green that labels the grass is the same green that labels the leaves or algae or seaweed. So when laws are agreed upon the same memory trace or very similar memory trace is brought into conscious awareness for all members of the group to consider as a base.

At the end of the day with law as the guidepost in a perfect world the goals of the group are identical to the goals of the individual COE. The priority in most basic form is the extension of life (longevity) and maintenance of homeostatic metabolic balance (happiness or security). Laws are human-fabricated tools designed to achieve those goals.

The threats to these goals generally stem from a rogue COE penetrating the physical energy perimeter of the individual human COE or by waveform energy causing a neural reaction. An imaginary threat however might do the same. In either case longevity or homeostatic balance suffers and laws are intentioned to curtail those conditions,

For example if it has been noted that people walking on a particular frozen lake have fallen thru on occasion it is because of an identifiable reason or reasons, one of which may be vested in delusion. The perceived density, quantity and arrangement of the supporting energy (the frozen lake) is a delusion.

A delusion is created in a number of ways: there may be a perceptual inaccuracy, there may be an inaccuracy in memory, there may be an inaccuracy in both or it may be a case of total ignorance. Total ignorance is the result of either or both lack of experience or absence of learning by observation of real and imagined situations.

Aside from a delusion of first-order there are also two other possibilities for the fall-thru occurrence, the first of which is *gambling*. In the case of gambling the human COE who walked out on the ice did have accurate memory and accurate perception but did override the safe response to chance the journey. The reasons for the gamble might range from no other choice, as if being chased by a bear, or a learned behavior of testing reality in hopes of a supernatural or miraculous result.

As strange as it seems, if probabilities in the past have been successfully defied by the human COE's decisions he may have developed a second-order delusion about his own personal energy construct within the ambient energy matrix. In this particular delusion the COE believes he has an unexplainable connection with energy forces which either he controls or act independently to help him achieve his goals.

Control of energy by humans will be addressed later as will the subject of God who is usually attributed to the later belief. It should be noted at this point the existence of God is not implied as a delusion in itself but can be instrumental in creating one by the human COE in question.

Back to the ice on the lake issue and laws. In any event and for whatever reason it has been observed by the group that individual COEs have either permanently dropped out of the visual field as a hole in the lake appeared or COEs have ceased all metabolic processes (died) after being retrieved from the lake or COEs have been observed in homeostatic imbalance after falling thru the ice. In order to maintain the goals of longevity and homeostatic balance for all individuals in the group, group memory and subsequent energy arrangement have been instrumental in developing a law to achieve those ends.

Laws have their origins in the memory traces of the individuals within the group. Those primary memory traces entail images in the examples of the above ice-lake tragedies or similar disturbances. The memory traces themselves were of sufficient quantities and magnitudes to cause homeostatic imbalances in multiple members of the group effecting symbolic communication thru words. The word symbols evoked imaginary scenarios in which an energy construct if effected by the group would prevent future tragedies on the ice. As different scenarios of prevention are hypothetically arranged in memory, the best ones are chosen as candidates for law. Those imaginary scenarios were tested by way of agreement with respect to accuracy of all the memory traces of the individuals in the group. If there was no disagreement as to the mechanics of the proposed, preventive energy construct the next step would be a neuromuscular energy response to produce written symbols (words) in the form of a universal instruction for all to perceive by way of light incidence on contrasting ink/paper media.

The law would have two parts. The first being a direct instruction and the second being a consequence for not following the instruction. In the case of the frozen lake the instruction would be "do not to cross the perimeter of the lake when frozen." The consequence for ignoring the instruction would then be stated. The consequence in most domestic cases is simply something the group will force upon the offender which causes a temporary, mild homeostatic imbalance. It could be a demand for material possessions such as money (to be discussed later) which will cause a condition of loss at the psychological level. Another consequence might be a limitation in mobility within the ambient energy matrix like incarceration.

All of these are intended to create an imaginary condition of proposed homeostatic imbalance *if the law was* not *to be followed in the future.* The mechanism of action relies on the individual's memory trace and mental imagery. In the majority cases those mental functions will create a simulated reality on an imaginary plane convincing enough to cause appropriate actions when the frozen

53

lake is within mobility range. The majority will walk around the lake rather than cross it.

There are those however within a minority who do not produce a negative-consequence-image real enough to effect a decision to follow the law's instruction. In that case the law will be ignored and if the subject does not fall thru the ice he might be penalized if there is visual or material evidence to support this history of his behavior.

Laws generally do not have a time span but might be changed or discontinued if not effective at enhancing goal attainment for the group of individuals it was created for.

The basic mechanics in light of pure physics are not that complicated; simply applying arrangements of energy to control other energy arrangements. Human COEs however get into all kinds of complicated issues with not only the creation of laws but the enforcement of laws and most importantly the contesting of those laws by members of the group. The biggest stumbling block in all of these issues is the concept of fairness or justice or equity.

If you look up the definition of any of these or like terms as fairness, justice or equity you will find it defined by the other two or three in the group. The most appropriate way to define these concepts in light of pure physics is a consideration of human COEs position in the energy matrix. If you position the human COEs as objective entities with respect to the ambient energy matrix (environment) and independent although identically subjected to man-made energy stipulations and arrangements (laws) conditions of equal bias or non-bias will prevail. This is understood as justice or fairness or equality.

The most prevalent complaint concerning the inequality for bias across the group, is that a law for one human COE is *applied* differently to another COE. This is usually evidenced in a perceived behavioral attitude which interprets the enforcement to be more constricting for one human COE as opposed to another.

If the mobility or opportunities to interact with other COEs or the ambient energy matrix are constricted it results in a disturbance of homeostatic balance. In some cases however it may actually impact longevity negatively. For example if in one position in the ambient energy matrix like California the human COE is required by law to be in existence for 18 earth revolutions around the sun in order to purchase a mental energy catalyst (tobacco) and in another position of the matrix like Alabama the requirement by law is 16 earth revolutions it could be a point of contention. The argument being the law in California is more conducive to extending longevity than Alabama by virtue of the fact that tobacco has a negative impact on metabolic process for all human COEs the same way.

The reason for the discrepancy stems from the lack of adherence to the original

54

in agreement qualifier of a law. Somewhere along the line the individuals who are contesting the law were either ignorant of the effects of two years of tobacco use on longevity when they agreed to the law or they had become aware only after becoming aware of California's evaluation of age requirement. So group awareness is key to the creation and functioning of laws.

In any event laws are not perfect and do not work one-hundred percent of the time. They are however arrangements of energy in a generally agreed upon configuration for the purpose of enhancing the longevity of a specified group or creating protection against homeostatic imbalance.

Up
And down
And in the end it's only round 'n round
Haven't you heard it's a battle of words
The poster bearer cried

Pink Floyd

Chapter Ten

Values, Quality, Good, Bad, Positive, Negative and Attraction

All of these things are somewhat related in both pure physics and everyday life. If you look up the definitions they all kind of intermingle and allude to the same family of concepts.

Except in specific cases where a negative result is desired, like a medical diagnosis or in electronics where the negative pole actually maintains an abundance of electrons compared to positive, all of the above describe some kind of polarization in which the *positive* is desired and *negative* is not. The good and positive adding to value and quality creating attraction and the bad and negative detracting from value and quality adding to a state of non-attraction. Even the word *attraction* has its origins in ancient medications such that when applied to a wound would draw out or *attract* infected tissue. All of this is central to an understanding of basic physics and pure physics as well.

At the core of pure physics is energy. At the core of energy is some kind of polarization. The engine driving polarization is attraction and non-attraction of whatever stuff energy is constructed of. In all likelihood it's just more energy in a smaller quantity and arrangement like everything else.

So if we choose to start at some imperceptible, immeasurable level of creation we still have polarized energy. If we follow that up the building-block chain thru pre-sub-atomic structures, to sub-atomic structures like electrons, protons and neutrons, to atoms and molecules and molecular conglomerates we have matter. If those molecular conglomerates had branched-off into DNA-specific conglomerates we have life and ultimately man, a human COE.

For whatever reason which may actually be symbolic, the intuitive knowledge human beings pivot their existence around has at its core those same rudimentary principles of polarization and attraction. Granted that in some cases the function of attraction is grounded in those same basic goals of extending life and homeostatic balance (for an actual reason) but some of the implied posturing is based on a more advanced understanding of values and reflection of those values as the center of existence. For example basic nutrient extraction from the environment.

From the instinctual get-go a condition of homeostatic imbalance might be experienced by a predator indicating the need for the replenishment of energy laden nutrients. Crossing his visual path light incidence causes a distinctive shape to create an image in his brain of a possible target to fill that need. The predator has a direction as he keeps that image within his visual field. When he takes that first step to create a larger image in his brain he is not only obeying a hard-wired instinct but also some law of attraction. He is reducing the ambient energy matrix between himself (the predator COE) and the hunted (target COE). As he reduces the distance between them a scent in the form of a molecular arrangement from the target might be inhaled to create an even greater state of homeostatic imbalance as "enhanced" hunger. That imbalance sends more energy to neuromuscular centers to further reduce distance at a greater rate. As the visual image gets larger it creates a state of tension in which the memory of the predator's lunge-distance capability is compared to the actual physical space between him and the target. If it matches he has a meal. In essence the attraction between two COEs has resulted in a synthesis of energy as the target is ingested.

The above example is the macro version of energy synthesis but none-the-less the same as on the sub-atomic level. It is further evidence that all existence is comprised of energy. The only difference in every construct being quantity and arrangement.

In the case of reproduction the process is similar except that the attractive forces are mutual to both parties and the assimilation ultimately ends up extracting more ambient energy from the environment to create a third COE. It is interesting to note human COEs readily enhance their outward appearance to increase attractiveness for mating purposes and this is a huge industry in human reality. Lower forms rely on natural physical features for the same purpose.

In addition to the biological or instinctual level of operation human COEs also entertain attractive forces at the imaginary level in an effort to fill an imaginary need. This might seem trivial but the end result most of the time concludes with a shift in energy or an energy-effected neuromuscular response which will satisfy that imaginary need. This usually results in the mobility of some COE in the real, tangible world.

Wherever there is a polarization of energy one side has the attractive property and the other side is either neutral or has an   abundance of whatever the other is lacking. This also holds true for a state of tension on the imaginary plane.

With respect to the aspects of *value* the motivation is similar. For example if a glass of wine spills on to a rug and causes a stain it causes no direct biological or instinctual alarm unless of course that wine was desperately needed for its nutritional value. At the imaginary level however a different scenario is played out. The rug has lost its *quality* or *purity* or *value* as evidenced by a light incidence difference from the pre-spill condition. Even tactile sensation may be impacted by the sense of wetness. Perhaps even a difference in odor may be detected after the spill. These new material conditions have labeled the rug in a negative way as "soiled or damaged." This reduces basic attraction by a human COE because of a deviation from *purity*.

Quality and value have similar vested properties of purity, completeness or goodness. All of these words sort of define each other. In the above example it all boils down to the rug as being purchased new and reflecting one certain pigment under normal, visible light up to and including its energy defined perimeter. Therefore it had the property of purity and completeness with respect to light incidence and reflectance when it was new or for a time after.

When the wine spilled on the rug it set that particular property of purity into an imbalance by virtue of contrast in reflectance. The sudden lack of purity creates a mild homeostatic imbalance in the human COE because it detracts from a *pattern of regularity* which is an innate measuring stick for security and reliability.

In an effort to restore that homeostatic balance the mechanism of imagination sets up a hypothetical energy trace based on memory which "attracts" a cleaning solution or cleaning condition to the rug magically restoring it to original purity. The image of original purity is accessed via another memory trace at this time for comparison as a model.

Further memory traces are subsequently called upon as possible sources for the cleaning solution or rug cleaning service. This results in a neuromuscular response to bring the rug to the needed repair application. This is in essence attraction by proxy. Although the rug and cleaning solution were not actually pulled to each other molecularly (although at the atomic level this will occur chemically in the cleaning process) they were indirectly manipulated by a COE to restore balance by the action of attraction on a macro level.

It can be argued that a rug free from stains is subjective as far as quality or value is concerned but you must remember human reality is an agreed upon construct based not only on individual perception but group decision. The individual may decide to override the desired condition of goodness or purity but in the majority of cases this will not happen. The net result is a condition of attraction to restore homeostatic balance.

The concepts of *positivity* and *negativity* follow a similar impact in the human real world. Positivity is a presence, an abundance, a completeness as opposed to negativity a void, a vacuum, a hollow, an absence or a need. It is for this reason positive is often aligned with *good* and negative with *bad*. The word *positive* is somewhat related in origin to the word *law*. Both are defined as a "laying down" or a laid down construct which serves as a demarcation. In the vastness of an ambient energy matrix a demarcation, line or sign post gives direction and may be a tool to define reality. A law is something not to be transgressed even on an imaginary plane.

The interesting part is the word positive is used in this sense to establish "something" as in something tangible, something real, something which is a construct of energy in the real world. Even if it is a barrier or restriction it is there not to be discounted. Negative is the opposite, a void of nothingness. So when humans use the word "positive" as in being certain or convicted about whatever premise is being referenced they are actually conveying a confirmation of reality. The assumption is that the larger group is in agreement on that point of reference.

Once again reality is being established by group conviction unless of course the individual was blatantly wrong. In that particular case a subjective delusion may be at the core.

Value and goodness as understood in human reality have a property of attraction to human COEs. An automobile (non-living matter COE) for example, might be attractive to a human COE because a memory trace suggests it can transport the human COE thru the ambient energy matrix to make possible the proximity of nutrients for extraction or other homeostatic-balance requirements. Advertisers will use light and sound to create images of homeostatic balance within "models or actors" who are staged as possessing their particular automobile product.

Those images are transposed thru sight and hearing to the viewers' memory banks. When any homeostatic balance is experienced by the viewers at a later point in time a memory scan will pull up possible corrective actions. Keyed to those corrective actions may be an image in which an arbitrary human COE (the model or actor), or even a self-image, is in proximity to the solution (owning the new car).

Having any car, let alone that particular manufacturer's brand might have no actual impact at all on the viewer's present state of homeostatic imbalance but the illusion was established. If the image is presented on multiple occasions a delusional memory trace may be established thru a process called learning. That delusion may create a chain of energy movements strong enough to create an attraction between the viewer (human COE) and the car (non-living matter COE). What follows would be the same series of events drawing the desired COE in close enough proximity to be possessed.

In similar words the human COE is the negative entity, lacking and in need, and the automobile a positive, good COE which when combined will make the situation a quality balance of completeness.

The concept of quality as cited in the earlier example of the spilled wine also exploits a quasi-delusional memory trace. Quality falls in line with goodness, completeness, positivity and high value. When a COE is labeled by a human COE of having quality or being of quality or requiring quality it is being called out to have properties of an imagined model.

A quality lawn mower is one in which the memory trace of an idealized lawn mower matches that unit. A high quality sales pitch is one that includes all those facets of memory which in the past has sold the product. A good quality car-waxing matches the imagined, shiny polished vehicle in memory. So quality like poles of positivity and goodness are standards for human reality to approximate as close as possible but they in themselves only exist in an imagined memory trace.

Human reality is again being created from a plane of non-reality or idealized imagination. If you add up all of the imagined intentions that are pivotal in making decisions for intended creation and compare those by either quantity or time to the actual constructs of reality anywhere in the present, you will find that the majority of time human reality exists on that imagined plane and dwarfs what is actually real.

It is for this reason that human progress as well as peaceful co-existence is often hampered. Everyone wants to create a reality with a face of goodness, positivity and quality but the seeds of implementation are vested in an imaginary plane. To complicate things even more, that imaginary plane is different for every human being. When one group's estimation of a reality that they wish to create is not aligned with another group's approximation the result is conflict and may ultimately result in war.

The concept of value is in line with the other ideas of measured goodness or correctness but has a curious twist which is not only present on the imaginary plane but is also vested in the very real condition of equivalence. The condition of equivalence is measured in the real world of tangible examples which do not change. It is for this reason that value only operates in the present. Ultimately value is metered in work and work is a product of energy transference or transposing. So value comes down to energy equivalence.

If an automobile has a determined value of $24,000.00 (currency and wealth to be addressed later) it means that a specific amount of work would be necessary to achieve that level of goodness or correctness as compared to the same relative amount of work to achieve some other end like a $24,00.00 luxury cruise. In the case of rarity the same logic applies. If a gem is valued at $10,000,000.00 in order to achieve that same level of energy transference that much work would need to be done, like the building of a mansion. The rarity of the gem and the rarity of the mansion are equivalent though different.

Value is also a product of imaginary transference of energy in the individual as well as the group. If the group does not have a unified, agreed upon work-equivalence for any particular COE or potential energy transference, the value is subjective and subject to speculation. If a rare coin turns up in a remote village where coconuts are traded for fish the coin is worthless. As for why a group of individuals would attribute a value higher than a direct practical amount to an item like a coin or painting or gem is to be addressed along with wealth and currency in another section.

Value is also metered along the lines of practicality or functionality to achieve some end. In this case value and quality are closely related at the imaginary level but at the tangible level something of value will transfer energy or do work in a direct and expeditious manner without speculation. A valuable ink pen will transfer ink molecules to paper molecules in exact configuration as intended via a memory trace and neuromuscular execution with certainty and without hesitation.

Hesitation is energy transference without regularity creating periods of non-transference by speculation. This is commonly referred to as wasting time even though time itself does actually exist.

actual matter and energy arrangements. Those items are labeled currency or money. Currency and money may further represent yet another symbolic group called *assets*.

Words and Definitions: ASSETS: a useful or valuable thing, person, or quality.

Property owned by a person or company, regarded as having value and available to meet debts, commitments, or legacies.

These are the actual COE's as owned, but the symbolic representation of these is a legally binding representation of ownership. What that means is that a group designated authority has enacted an ink-on-paper light-energy contrast which stems from symbols in memory on an imaginary plane. That image in memory is agreed to by the larger group of human COEs to be identical to that of the individual members' memory traces of that group.

One additional symbol of ink-paper contrast represents which member or members of the larger group have ownership to that asset (identity).

Now it is possible for that particular member to "attach" that symbolic asset to the outside of his energy tube albeit on an imaginary plane. If he so chooses he can trade that asset for currency or other assets. He can also attract the physical matter represented by that asset symbol to himself if he so chooses. This is only if the asset has some intrinsic value based on proximity to the owner.

If the asset is a boat he may want to use the boat to transport him in the ambient energy matrix to be able to attract COE nutrients that exist in the sea. He may just want to propel himself and other human COEs across a liquid energy matrix for recreational purposes (recreation to be discussed later).

Lastly the asset may be used to collect additional currency by virtue of an agreement with other human COEs in which currency is exchanged for work facilitated thru the asset which he owns. Work being a transition or transposition of energy being facilitated thru the asset for transposition into some other exchangeable COE.

Both currency and assets are desired COEs even at the symbolic level. The accumulation of these exceeding a group-determined assessment of quantity value is labeled as *wealth*.

Words and Definitions: WEALTH: an abundance of valuable possessions or money.

The derivation of the word however is more significant. Wealth is derived from old English as *well* as in *well-being* and later as *health*. This is important because other than an entertaining the current materialistic place in the scope of human reality wealth was a skewed interpretation of health and well-being. This is important to understanding human reality because health and well-being are linked to the basic goals of all human COE's: the continuance of metabolic processes (life) and the maintenance of a homeostatic balance. It is probably for this reason that wealth is a major impetus for the transference and transposition of all types of material COEs and non-tangible energies.

Many humans are so swept up in the acquisition of currency in hopes of reaching the point of wealth they spend the majority of their active waking hours striving to achieve that goal.

The escape from normal logic on this behavior is supported by a two-fold reason. First and foremost once past a certain point in currency and/or asset accumulation any further amount traded for work or energy will not further increase metabolic longevity nor create a more homeostatic balance. It should be noted that this point may or may not exceed the point labeled as wealth by the larger group. So any further accumulation is either a lark or a misconception about wealth's life enhancing properties. This is in essence a created delusion.

The other aspect concerning the over-accumulation of currency or assets culminating in wealth is that the particular delusion this condition is driven by is "attached" to an ancillary delusion. That ancillary delusion suggests that the asset quantity which is enough to guarantee the maximum possible metabolic longevity and homeostatic balance *is always slightly more* than the actual quantity of assets accumulated *at any point* in time. In fact the word asset has as its origin a translation of the word "enough." Surprisingly these two delusions are strong enough to cause many human COEs to cross both legal and safe limits of active energy control in pursuit of wealth.

The mechanics behind these delusions all stem from errors of memory, perception and ignorance (lack of previously proven knowledge). It is also surprising that so many human COEs not only use most of their active hours trying to satisfy the delusions but will actually put real, necessary life sustaining constructs as a second priority in pursuit of those fantasies. Sometimes those fantasies are purposely promulgated by humans themselves.

Humans like lower primates will imitate a perceived action if it appears to benefit the actor. If a human COE witnesses another COE in a state of homeostatic balance and an extraneous, though proximal COE is attributed to that state of balance, the observing human will attempt to attain that extraneous COE in hopes of creating that same state of balance in himself. Since the energy transposition at its origin takes place with perception, the credibility of the cause-and-effect, (i.e. the extraneous COE actually producing homeostasis) is limited to the perceptual

65

field of the observer. What this means is that the cause-and-effect can be credible or "real" or it can al be staged in such a way so that normal limits of perception would render the exact same energy transposition for immediate memory as a "real" condition. The latter "staged" condition is the object of commercial advertising and is effective in creating a state of attraction between the human COE and the extraneous COE. That extraneous COE being an asset by the way can usually be exchanged for currency and this is the intent of the whole thing.

The attraction is created by a comparison in memory of the observer in which the perceived actor and a memory trace image of the observer are arranged in association. On the imaginary plane the actor and the observer are equal with the exception of the ancillary COE and the appearance of homeostasis. This creates a mild homeostatic imbalance in the observer which when coaxed by the ad is attributable to a lack of the ancillary COE. The resulting condition is labeled as *desire*.

Desire is similar to a need but is more elective and subjective. None-the-less desire can propel a memory trace to decisive action to attract that extraneous COE on a real plane of tangible matter. This almost always involves a trade of symbolic value (currency) for the desired COE.

Even before mass media took control of commercial advertising the condition of created or invented desire existed. As long as another human had some COE "attached" to his outer tube and an observer did not, the observer would experience the need to attract it (desire). This is a throw-back to the original premise of energy with respect positivity (something) and negativity (nothing).

As communication across the board developed and was perfected more and more created desires could be entered in the human memory bank. With an overflow of images desires are easily confused with genuine needs. This resulted In a manufactured condition of homeostatic imbalance being created on a large scale. That particular homeostatic imbalance could once again only be corrected with action on the part of the human COE observer. The object of the action to restore a contented balance is effected thru an energy trade-off.

With all of the different desires coming to fruition the most practical method to deal with the vast variety is by way of an abundance of currency. So the goal to attract and accumulate currency or its liquid equivalents in hopes of a trade for all those desires accumulated in memory takes precedence. Although a fantasy this is the engine behind wealth accumulation behavior.

As a corollary if the action to accumulate wealth is so pronounced that the human COE is branded with the same behavioral intent even when desires have been satisfied on regular basis it's labeled as *greed*.

Significant currency culminating in wealth also allows for the control of extended COEs both human and non-human to some intended end to actually attract more currency. This is one of the ways more wealth can be accumulated over time. It is essentially an investment of energy arrangements which will attract more energy moving potential for the movement currency in some planned direction.

Human COEs are often labeled by the larger group as to the quantity of wealth within their control both in tangible matter, tangible currency and assets on an imaginary plane referenced by symbols on paper. A hierarchy based on these accumulated quantity differences was also established with the most accumulated wealth at the "top" and the least quantities of wealth at the "bottom". The concepts of top and bottom are of course also illusionary and only exist on the imaginary plane but are none-the-less considered one of man's biggest barometers of his goodness, quality, purity or excellence.

The reason for this stratification based on wealth is somewhat speculative but appears to be linked to the perceived quality of non-human COEs that can be "glued" to a human COE's energy tube at any particular time. It also refers to the perceived quality of energy movement or work that can be obtained for that higher quantity of currency or wealth.

Some of this is true to a point and some of this is posturing based once again on delusion or fantasy. When posturing is exaggerated it is labeled as pretentious or ostentatious and this also has a purpose. As advertising it indicates to other human COEs that this particular COE has the potential to move large amounts of energy which might be pooled with their own contributions of wealth to achieve some end like attracting even more currency and wealth.

Another reason wealth might be displayed or inadvertently "advertised" is to use it as a lure for other human COEs lower on the wealth scale. More often than not another delusion or illusion is created within the mind of the observer in which the displayer of wealth is of an empathetic nature and will transfer currency or some valued asset in trade for little or no work-energy. This attracts a number of human COEs as a method to accumulate wealth with little expenditure of energy. What usually happens is the display of easy wealth was staged at some level. The work-energy extracted from the observers is actually at a greater rate than presumed before he is the wiser.

A display of wealth also serves to keep the quality of proximal available COEs at a minimum desired level for the wealthy human COE to "glue" to his tube. This happens because he has *displayed the capability* to attach those COEs if he so desires.

Currency and wealth is also used as a tool for a group of human COEs to move energy in a way which will benefit the group. The pooling and collection of currency may be termed a tax or fund or collection. What is happening here is very simply an agreement between individual human COEs to pool their symbolized energy movement potentials (currency) which they have independently accumulated (probably by neuromuscular energy expenditure). The collective, larger potential can then be channeled thru some other conversion into work or COE transposition. The end result is a change in proximal energy constructs for the contributing parties which will benefit them in some way. Were it not for the pooling of currency the feat could not be accomplished.

A twist on this is the case of charity in which the pooled contribution of energy movement potential benefits an alternate group or individual human COE lower on the wealth scale than the contributing members. The reasons for this behavior range from a supernormal directive (God and religion, to be addressed later) to an implied return of the contribution at a later point in time to simply a learned memory trace without direct support.

One last note on currency, assets and wealth with respect to accumulation without apparent generation by work expenditure. This is termed either gambling or investment. Gambling is an agreed upon, pooled, contribution by a group or unitary human COEs in which that pooled contribution is attracted to the winner. The winner is a member of that group who correctly indicated an arrangement of material energy which all members of the group agree exists in that particular arrangement but don't actually know when. The winner must first create that arrangement on an imaginary plane and then thru light or sound waves communicate his guess at the arrangement to the group. That winning arrangement may be one of many energy arrangements at any point in time not directly predictable with certainty by any human COE. The fact that any human COE does not have the perceptual or memory trace capability to consistently and accurately imagine how the winning energy arrangement will be structured is the engine behind the action of gambling.

Why the human COE chooses to partake in this action is vested in either ignorance, a belief in a supernormal connection conveying the winning arranged, configuration to him alone or recreation (to be discussed later). In any case human COEs have both accumulated and disseminated currency thru gambling.

The other method is labeled investment. An investment is the allocation of currency from one human COE or group to another COE or group that has agreed to return the currency in a set time period at a greater amount. Investments can range from gambling at one extreme to a positive prediction of return at the other extreme.

Essentially the initial symbolized energy movement potential (currency) is used to move other energy into an arrangement to attract additional currency. This currency is then moved back to the original human COE investor. The only caveat to this is a time stipulation. The investor COE is only able to maintain his metabolic processes (life) for a finite number of energy transformations (time). If he needs to utilize the additional currency it needs to be returned in within that period.

It is worth mentioning that crime has also been used to accumulate wealth. Crime is the act of crossing real or symbolic barriers of both human and non-human COEs to attract currency or wealth. It is akin to gambling because more often than not human COEs who enforce those barriers attract any collected wealth back and impose a time-specific material barrier on the criminal to limit his physical mobility within the ambient energy matrix (incarceration).

Forward he cried from the rear
And the front rank died
And the general sat
And the lines on the map
Moved from side to side

Pink Floyd

Chapter Twelve

Governments and Administrations

Words and Definitions: GOVERN: The word *govern* stems from the Greek term meaning "to steer." In human reality governments and administrations in the ideal sense are supposed to function in that fashion. The actual physical reality is somewhat different. First off there is no actual steering of a country or municipality unless it is on the march to some other location. It sounds ridiculous but if you look at the issue and process thru pure physics you will see there is very little direction involved in governing although the terminology is used without thought.

Governments and administrations are closely aligned with extended groups of human COEs. The subject was purposely not addressed under the heading of *Family and Extended Groups* because an understanding of currency and material wealth is necessary to understand the workings of government at least in times of civilized man.

Before organized systems of currency came into being governing was enacted thru light and sound images in real-time as an instruction and consequence for not following that instruction, i.e. the use of force. The determination of what instructions were necessary to achieve some end was the province of the strongest of the group. The weaker of the group were the ones to effect a transposition of energy in an effort to achieve some goal as created on an imaginary plane by those governing. If this process was not carried out within a certain degree of accuracy those weaker individuals would be subject to inflicted homeostatic imbalance (pain or loss) or even immediate dissemination into the ambient energy environment (death).

It should be noted that some of those goals on an imaginary plane *may* have had some benefit to the group or had just been a frivolous construct concocted thru ignorance or arrogance. An example might be the building of a pyramid or slavery. As man became more civilized governments worked more towards the benefit of the entire group. This probably evolved because of the growth of formal education and naturally developed enlightenment. As the majority became more in-the-know individuals in groups could manipulate a governance in their favor by shear quantity vote of the correct interpretation of reality. This was supported by the increased amount of potential energy transposition now within their reach.

In modern times that potential energy to transpose other energies has become standardized currency and material wealth. This is a two-edged sword however because governments also use currency and wealth as a tool to accomplish some end which may not necessarily favor a group benefit.

The best place to start with the subjects of governing and administration is with extended groups beyond family. It is true that some national governments are strongly vested in a family structure as a monarchy but even at that it's usually immaterial with respect to the actual mechanics behind group actions.

Essentially groups of human COEs will come together in ambient space because individual human COEs are influenced by mutual attraction based on needs. Positive aspects of attraction might be for regular nutrient requirements or some transference of energy in material forms like chemical or intangible forms like electricity.

Within groups there are almost always energy transpositions from energy to work and work to energy. As stated previously these actions are almost always intended to extend metabolic processes (life) as far as possible into the future or maintain homeostatic balance for each individual member of the group in the present or anticipated immediate future.

When human groups were tribal and in small numbers the leader of the group may have been pivotal in actually "steering" the group towards a nutrient source or away from an enemy and this was mostly determined by the input from various members. If the leader was the "head" of the group the group itself was the "neck." The determination as to who was the leader may have been by group design later but early groups did revert to dominance by the fittest, the strongest, the smartest or the most violent. The important part at this point in time was the group's input in "steering" the group while using the leader as a signpost for the group's decision.

The leader was also the word of law in that whatever that signpost had scripted each individual of the group needed to obey. Primitive government was born and took action.

If we pick a hypothetical, pure physics, starting point what was occurring was a homeostatic imbalance within one or more individuals of the group which may or may not have been warranted. Based on the early time-line however primitive man usually took action out of necessity for survival or a perceived necessity. That imbalance resulted in a neuromuscular response in those affected members to mediate light or sound waves to other members communicating the alarm. The image created in memory trace on an imaginary plane for those other members may have been sufficient to create an additional though imagined homeostatic imbalance within the group as a whole. This in essence is a form of group agreement as to a universal threat or need.

If a sufficient number of members were prodded to that state of imbalance they would have effected a neuromuscular response in the form of light or sound waves to the leader. The leader if functioning as a proper leader would have taken action by his own neuromuscular response in light and sound to echo a focus on the issue.

As a rider to the original declaration of imbalance by the group was probably a solution or multiple solutions for corrective action. On an imaginary plane this served to temporarily restore homeostasis. (This aspect was not mentioned earlier in the scenario because it would have really complicated things.) In any event the imbalance and the method to correct the imbalance are echoed back to the group members from the leader as an instruction which each member must heed or suffer some implied consequences.

That is the rudimentary basis of modern government with the addition of some twists and turns which were necessary due to the increase in group size and

subsequent increase in government size.

One of the biggest differences between tribal government and modern government is the use of money as a symbol of potential energy transposition. Another difference includes an elected body of advisors and mediators to facilitate decision making and communicate needs.

Starting from square-one as in the tribal example, the larger, extended group has needs both necessary for survival and for a pleasurable existence free from worry. Some of those needs need to be mediated thru laws and others as the enactment of work and material energy transposition in matter. Because the size of the groups has grown upward of millions the task of communicating needs becomes more complicated than simply a shouting leader. Also the tasks have become more complicated than simply moving a hunting party in the direction of game. In fact very little mobility of large groups happens with the exception of a natural disaster or war (war to be discussed later). Most of the modern day "steering" by administration is in the movement of energy proximal to the human group to achieve some end.

In modern times a group need follows a similar path as the tribal process. Because of the increase in size however that need warrants a number of individual members of the group to effect a light incident or sound wave alert to any one of a number of centers for amplification. Those centers might be the news media which will multiply the message in distance so a greater number of group members are alerted to the condition of homeostatic imbalance experienced by some individuals.

An administrative sub-group of advisors may be subjected to this information by way of the news or may in fact be signaled directly by the individuals in need.

Various administrative sub-groups at different levels may serve the same general purpose of a communicative conduit. At some point a decision-making facet within that initial sub-group or an ancillary sub-group will use that information on an imaginary plane to gauge the priority of the imbalance. This happens because with large groups multiple imbalances are evident and communicated on a regular basis at any point in time.

If an imbalance needs to be corrected with immediacy the need on an imaginary plane is communicated across the advisory group until it is validated as the same memory trace in each individual. Once this is established corrective action solutions are compared by first validating the identical solutions on that same imaginary plane for each individual in the advisory group. What follows is either signaling agreement or disagreement between the members. If the signal is one of agreement the proposed corrective action is labeled as viably effective. Next is the agreement of a theoretical time frame for the taking of the imaged memory trace to use as a model for moving actual energy. Once this is established the usual course of events is to move a symbol of energy transference or transposition (currency) to specialized human COE receptors. Those receptors will then signal via light incidence or sound waves to human COEs who will either execute neuromuscular action to expend energy (work) or effect a mobilization of non-human COEs (matter) to a location proximal to the original need. In any event these actions will ideally restore a homeostatic balance to the overall group within the expected time frame including those individuals who initiated the action in the first place.

There are a few more characteristics central to the modern, working human government as remnants of primitive man. First there is still a leader or main human COE who still serves the same basic purpose. He is a figure head and signpost for the current energy transpositions of the group and imaginary or proposed transpositions time stamped for the future.

His role has expanded however to also include a capacity of attraction for other group leaders of similar elevated positions. The reason for this is to ensure the mutual attraction of both groups' individual members on an imaginary plane. The mutual attraction imagined is one which will promote the extension of metabolic processes in time (promote life) and maintain homeostatic balance for all members of both groups.

Generally a mutual transference of energy in both matter and work between the two large groups, contributes to this congeniality and is referred to as trade. The whole condition between the large groups is labeled as peace as opposed to war.

If the constructs on the imaginary plane for both group leaders is in cadence it is considered beneficial for both groups. If there is a discrepancy in image and memory trace between leaders which cannot be settled it is probably attributable to a philosophical difference resulting in a difference in attitude. What this means is that the imaginary plane for one or both leaders is not supported by tangible or proven counterparts in present reality and is instead based in intangible belief which cannot be proven. The net result is a mutually detrimental relationship between the two groups.

Other facts about modern government include the positioning of specific human COEs in various specialized energy expenditure (work) roles within the government structure. They are selected by group assessment as to quality, pureness or goodness in their energy transpositions. Selection is also based on the individual's past history and self-testament of as to energy transference and transposition capabilities. All of this is evaluated by the larger group before seating the individual in a specific position in government.

There are also sub-groups within the government which test the actions of key members. This is done by comparing those actions on an imaginary plane with logical known energy constructs as effected by those members. If the actions balance out in accordance with those constructs the member is said to be credible. If there is a discrepancy the member may be inaccurate in his arrangement of memory trace or having had experienced a perceptual error causing the invalid action. The perceptual error may stem from another human COE conveying misinformation that was also inaccurate thereby creating a chain of errors. If too many alarms go off to that effect the member is usually replaced with an understandably more reliable candidate.

The cross-over from tribal leadership to geographical group position leadership along with the development of currency and the concept of material wealth greatly impacted the workings of government. Currency and material wealth became one of the biggest factors determining the effectiveness of administration. This became increasingly more important as civilization developed even into the present day.

Thru the use of currency there was less and less physical mobility of governmental influence such as dispatching a legion of troops to raid a village. Currency became equivalent to electrical energy in an electronic circuit. Currency is in effect the potential to expend energy or transpose energy of many forms at a distance. By using currency a government can tax, or attract and accumulate energy potential, from a village without actually dispatching troops. That energy potential then pooled by government can be then used for the benefit of the group.

Although problem solving ability and the articulation of options to solve those problems is boasted by modern government systems, the real power to move energy is still vested in that government's pooled energy resources and material wealth. The difference between desired administrative systems of government and undesirable ones is usually paralleled by the governments' effectiveness in its material wealth and currency handling system.

Because the ability to move energy by currency is based on quantity extracted and accumulated from the individuals in the larger group it limits the individuals' capabilities to satisfy basic needs directly at their level. If a fair or equitable balance is kept between the government's share and the individuals' shares a homeostatic balance is maintained for all parties and it is considered just.

How a government came into being is equally as important. If the government was established without the consent of the group it is a good indication that it came into being by *force.* This is essentially a light and soundwave incident which creates a homeostatic imbalance in the larger group by a smaller invading group. The imbalance is usually first created on an imaginary plane in which dissimilation of the larger group is imagined at the suggestion of the invading smaller group by way of COEs (weapons) in their possession which when effected by neuromuscular action will cause a cessation of metabolic processes (death).

If the larger group does not endorse the invading smaller group at this point in time one of two things can happen: if the larger group also has COEs (weapons) capable of dissimilating the smaller group it may set up a homeostatic imbalance great enough in the smaller invading group enabling them to be attracted to another position in the ambient energy environment less proximal to the larger group (retreat). If on the other hand the larger group does not have the COE resources (defense) to cause that imbalance, the smaller group will probably cause some dissimilation in the larger group causing further homeostatic imbalance which would need to be corrected by compliance of the larger group. This condition of government is labeled a *dictatorship by force.*

Dictatorships are a form of government in which force is the usual method to get things done benefitting the government first and the individuals in the larger group second. The engines behind dictatorships more often than not are ignorance and delusion vested once again in material wealth.

In conclusion it should be noted that governments in the ideal sense also function as a measuring stick for the quality of most energy transfers and transpositions within the defined physical perimeter of the group. That might be city, state, territory or country. The reference to an ideal level of operation is necessary because much of the intended, effected actions of governments have their origins on an imaginary plane of idealism or perfection.

It might be added that the images and memory traces of one sub-group of government is often blurred in comparison to another sub-group of government.

This is also the case between a government sub-group and the larger group of governed individuals. This condition is unavoidable considering the massive quantities of images and memory traces at any point in time and the constant changing of those images over periods of time. This creates discrepancies for anticipated response actions as well as the interpretations of the actions by the larger group.

It is for this reason most members of government as well as the larger group spend a great deal of time comparing ideas and arguing over those ideas and what the imagined effect would be for the larger group of human COEs. The criticism of inefficiency is often voiced by the larger group with respect to administrative operations for this reason.

Ooh, see the fire is sweepin'

Our very street today
Burns like a red coal carpet
Mad bull lost its way

War, children, it's just a shot away
It's just a shot away
War, children, it's just a shot away
It's just a shot away

Rolling Stones

Chapter Thirteen

War and Peace

Words and Definitions: WAR : a state of armed conflict between different nations or states or different groups within a nation or state. Word origin is from old Anglo-French-German *guerre* which stems from *worse*.

The origin of worse is easy to see because since the beginning of the existence of humanity there has existed war and it was always considered one of the worst outcomes of group interaction.

The word *conflict* is key here as well. Conflict is a disagreement or lack of approval over some issue in which one group does approve.

Words and Definitions: PEACE: freedom from disturbance; tranquility. A state in which no war exists.

The key word here is *from*: Indicating the point in time at which a particular process, event, or activity starts. Indicating the point in space at which a journey, motion, or action starts.

The word *from* also means to a lesser extent: Indicating prevention. Indicating separation or removal.
Indicating a distinction.

All of this word stuff will come into play later but just to clarify *armed* is understood to mean having a weapon in possession. In the world of pure physics armed means a human COE with an attracted and "attached" non-human COE of higher density than the maximum possible human COE density.

The non-human COE is considered a weapon if *weapon* was a previously agreed upon label and exists in an agreed upon imaginary plane as such or if it was actually used with intent as a weapon in the real tangible world. In either case the non-human COE weapon would require the addition of energy either directly or indirectly from a human COE to be effective against some opposition.

Neuromuscular energy directly as with use of a sword or indirectly as a catalyst to a charge in a gun is activated to full potential by a chain reaction effected by a human COE. The end result is the same if implemented against another living COE but just for the record: the high density non-human COE when put into action invades the perimeter energy construct of the target causing a disruption in metabolic processes sufficient to either limit normal COE neuromuscular movement or ultimately terminate metabolic processes completely initiating the dissemination of energy into the ambient energy environment (death).

As far as human reality and pure physics are concerned war itself is an escalated chain reaction not entirely different from a nuclear chain reaction. In fact this might the logic behind the nonsensical notion of nuclear annihilation. Instead of reality following physics, physics is put into effect to create some imbalanced individual's idea of what reality should be.

The war-reaction or condition starts with a homeostatic imbalance in individual, human COE members of a group. That homeostatic imbalance is registered in the memory banks of the individuals experiencing it as significant. By way of light incidence and sound waves those memory traces of the individuals are cross-checked within the group for an accurate similarity. If found to be identical on an imaginary plane the group will seek to identify and label a cause. That cause is often construed to be the same across the board even though there might be different causes for the same effect experienced. If the identified, labeled cause can be attributed to another, alternate group of human COEs war may be precipitated.

It should be noted that the group thought attributable to the cause will need to be clearly distinguished from the accusing group. This difference in identity may also contribute to causation but might also be irrelevant.

Some of the reasons for the homeostatic imbalance causing the reaction in the first place may be a real or perceived inequity in the possession of tangible COEs claimed by the two groups. Things like nutrients and potential energy stores such as crude oil might be of a greater quantity as possessed by one group as opposed to the other. Another reason may be attributed to actual physical space within the ambient energy environment. Both groups might choose to define a physical space symbolically by way of contrasting ink on paper and label that as a possession. If both groups share the space no problem ensues. If they each wish to be in control of all COEs within that space including human COEs at a factor of 100% it is not possible. If they both cannot tolerate the condition with a compromise war is probable.

Another reason for the event of war might differences in political, social or religious ideology. For any number of reasons the ideologies of one group might be so self-vested in memory that any deviation from that ideology causes a homeostatic imbalance. The ideological memory trace might be so strong that a perceived deviation in other human COEs even at considerable distance causes an imbalance. In any event the important points are an imbalance and attribution of that imbalance to a labeled dissimilar group.

Once demarcation takes place on an imaginary level one or both groups will begin to act on those memory traces in an effort to restore balance. The usual corrective action chosen is one of causing a greater homeostatic imbalance in the opposing group often at the expense of cessation of metabolic processes (death) in some of the opposing group's members. In genuine war this leads to a similar polarized response towards the group initiating the war. This instigation of response generally goes back-and-forth, each time further diminishing homeostatic balance in both groups. It usually also dissimilates both human COEs and tangible matter COEs into the ambient energy environment. After a period of escalation and plateau war finally ceases when one or both groups agree to some compromise because it will reverse the expanding homeostatic imbalance and limit the ongoing reduction of living human COEs thru dissimilation.

This state of achieved peace is characterized by a return to some sense of normalcy in that a homeostatic balance is returned to both sides. That state of peace may last days, months, years or centuries.

Lasting peace may be difficult because much of the time the memory traces of the past war gradually move from distant memory into present memory by either chance occurrence or by a deliberate effort by members of one group to "relive" the transgressions. Once the memory traces are in present memory they cross over into consciousness. Consciousness is generally reserved for immediate problem solving and behavior in the present or immediate future. If an emotional "tag" is attached to the ideas a homeostatic imbalance may be effected, albeit on an imaginary plane. From this point on a back-and-forth reactive spiral may ensue in repetition of the original impetus.

It's interesting to note that from the perspective of the present day human COE this understanding is not considered a major advance towards keeping lasting peace. He does not bring the reality of this relationship into consciousness for at least debate in hopes of eliminating the state of war altogether. Instead the focus is on proving the impetus for war is with the opposing group. This results in justifying an escalation as a "no alternative defense." Even more weight is given for the argument of the ever popular "preemptive strike."

The human creation driving conflict is a proposed logical construct which suggests the winner is justified in their actions because their argument was validated for goodness, quality, justness or equity and the opposing group's argument lacked those attributes. This *may be* grounded in reality (especially in Hollywood) but more often than not both groups will use this argument simultaneously.

If the pivot point of the war is one of material, tangible matter or a geo-physically defined space the concept of *entitlement* is the driving force. Entitlement suggests a "right" to something. The word right is from old Latin as "ruled" as in a straight line. This crosses over to correctness, quality and goodness most probably because anything that was straight was of value especially a path to food, water, etc.. The arrogance associated with those virtues translates into entitlement.

So from square-one if both groups are arguing entitlement to a coconut or whatever it is they are fighting over, each has a memory trace on an imaginary plane which supports this. When those memory traces are compared within the same group they are identical but when compared across groups, dissimilar.

The reason the memory traces do not coincide exactly is due to either ignorance or denial. Considering denial first because it is the easiest to understand, one group wishes to possess control of the coconut for whatever reason and chooses to justify their actions thru a self-imposed delusion. They know in fact they do not have entitlement but cannot on a conscious level accept the loss of the coconut. Some absurdity is fabricated and sworn to as a reason. Ignorance on the other hand is a little more complicated. The state of ignorance may exist on any one of a number of levels. At the very worst the level is at face value. This means the actual aggressor believes he is entitled because he likes the color green and the coconut is green. At deeper levels the absurdity is generated at an earlier point in time or by virtue of some other inaccessible, untraceable condition. It is then postured into a plausible issue and supported by quasi-facts that are hard to prove or disprove. This is then promulgated as a "factual" reality with fervor. The fervor leads to imbalance and the imbalance leads to aggressive action and so on and so forth until a state of war breaks out.

The basic desire in both of these conditions can stem from any one of many conditions of need real or imagined. Generally speaking when it comes to war an element of delusional greed enters the equation somewhere and is very similar to the conditions associated with currency and wealth.

Along with tangible material objects of desire other reasons war may be precipitated are centered around group-aligned differences in ideology. *Ideology* is defined as a system of ideas and ideals which form the basis for social, political, religious, philosophical policies. *Policy* is defined as a course of action as in civil administration. In any event all of the above subjects are debatable on an imaginary plane. So the short answer is wars are fought over the way things should be done within a group. Because almost exclusively anything can be done in a number of ways there exists extreme fertility for war.

From a perspective of pure physics, all human COEs within a group are subject to energy flow from other COEs both human and non-human. Some of these may be tangible matter like a box of potatoes. Some of these may be purely transient energy swings in the environment like heat energy or the lack of as cold. Some of these may be symbols of energy movement or potential movement or transposition like currency and material wealth. In order to create a predictable transposition of energy for the group an ideology or policy is enacted usually by the governing members. This ideology is a flow-chart or road map for relevant COEs so that when acted upon by individual group members it will be transposed where ideally it is most beneficial to the group with equal fairness to its members. This may be a compendium of laws or religious doctrine or a system of politics in which the control of energy is moderated thru an established chain of events or actions. This tends to give a quality of predictability to life in general thereby supporting homeostatic balance.

Because these laws, doctrines and systems are created by human COEs on an imaginary plane the same pitfalls arise as in the case of entitlement for material objects. The group aligns itself along agreed upon memory constructs that are supported by both fact and quasi-fact. To add to the fun whatever policy or law is being validated for "goodness, quality or rightness" may in fact be effective for whatever it is supposed to control but it is not the only game in town.

This later caveat is usually the reason which precipitates the conflict. It's basic common sense. There are always at least two ways to do things and pending subjective history one or the other is going to be determined as the most effective. The problem is subjective histories are as varied as snowflakes. Once group-think generates a state of arrogance and fervor the "tagging" of the opposing group as being a non-problem solver generates that initial state of homeostatic imbalance. From there on it's off to the races depending on level of conviction of each of the differing groups.

A large off-shoot is an attempt to justify the ideology argument for war by advocating freedom. Freedom is defined as the power to act, think or speak however one wishes without hindrance or restraint. This means a human COE can execute neuromuscular energy to effect physical movement or sound waves in any possible arrangement of energy within the ambient energy matrix. It also means the human COE can arrange memory traces in any possible arrangement or order of occurrence at any point in time. That is an extreme definition and when the symbolic word "freedom" is normally used it is understood as if some constraints are necessary as the law or social convention. This is further crossed over into an interpretation that means freedom from bondage or slavery.

If one group of human COEs is being dictated a course of neuromuscular actions that each member must effect under threat of a homeostatic imbalance or premature cessation of metabolic processes, it is labeled as slavery. Similarly if the arrangement of memory trace is limited as is the arrangement of emanating sound waves, freedoms of thought and speech are hindered. Varying degrees of these actions range from a label of oppression to actual slavery.

Although more prevalent in the early stages of civilization these events still take place in modern times however usually considered internationally illegal. As an outgrowth of tribal warfare and domination the winner sometimes found it was more advantageous to enslave a defeated enemy rather than annihilate him. A slave meant energy could be controlled with little expenditure therefore transposing a small amount of energy into a large amount of work.

A corollary to slavery is *persecution*. Persecution is a combination of war at face value and slavery. It is the intentional causation of homeostatic imbalance, sometimes to the point of pre-mature metabolic cessation based on some perceptible difference from one group to the other. It is the absurdity of war on steroids.

Freedom itself is somewhat oxymoronic in that wars are sometimes fought to achieve freedom and sometimes freedom is lost due to defeat in war. This is further supported by one definition of peace as *freedom from disturbance.*

There is nothing more disturbing than war. In fact this oxymoronic quality of the word symbol itself may be the engine behind the whole ridiculous action of war in the first place.

As far as the big picture in pure physics is concerned war can be defined as a *reactive synthesis.* The all-inclusive group labeled humanity has split into at least two distinctive, opposing groups based on attracted material wealth or attracted ideology on an imaginary energy plane. Because of these differences in proximal energy one group initiates a homeostatic imbalance in the other group which copies the action. This causes greater cohesion and internal attraction within each group (ask any soldier or country at war how they feel about their fellow countrymen and you will find no greater affinity). That attraction reaches a point of saturation where attraction to the opposing group is initiated and peace ensues.

It's the basic behavior of energy at a macro level.

Can you surry, can you surry
Surry down to a stoned soul picnic
Surry down to a stoned soul picnic
There'll be lots of time and wine
Red yellow honey, sassafras and moonshine
Red yellow honey
Sassafras and moonshine

The Fifth Dimension

Chapter Fourteen

Education, Recreation, Entertainment and Leisure

Human reality would not be complete without activities outside of those directly maintaining metabolic processes, transposing energy (work), replicating and maintaining basic homeostatic balance. Some of these other activities may not be necessary for survival in a direct way but contribute to that end by helping to extend longevity. All of these do have a highly subjective value depending on the development and memory trace arrangements of any particular human COE. It is thru this varied value and contrived necessity that extreme diversity in all of these areas is evident in human behavior.

Education is a condition of intentional or unintentional energy transfer in the form of memory in symbol and image. It may take place between two human COEs or a human COE and some other COE in tangible material form or a transient energy state. If it is only a one-sided human COE transfer the memory transfer is effected from real life energy as opposed to energy created on an imaginary plane.

If it is an intentional memory trace transfer the reason almost always is to facilitate the transposition of energy at some later point in time. That energy transposition may be an effort to recreate the original memory trace as accurately as possible as in the case of learning a musical arrangement. If not used for identical replication the established memory trace may be arranged as a usable construct on the imaginary plane to accomplish some other practical end. If the arrangement proves theoretically that an ancillary task can be completed in less time or with less energy expenditure the arrangement may be effected in the tangible, real world to that end. For instance if learning how to mow a lawn with a power mower is employed to save time instead of mowing manually takes place, the next time the lawn needs to be mowed a power mower will be used.

Human COEs generally place a high value on education and the process of being educated ranks high on the priority list as far as activities required during normal life are concerned. This is particularly pressed into service within the early growth years because memory traces can be more easily arranged when human COEs are young and they also remain as arranged for a longer period of time.

Another reason is that the earlier a human COE has a coordinated collection of memory traces within his memory banks the earlier he can effect the transference and transposition of energies efficiently and correctly in the ambient energy environment. It is for this reason that only useful arrangements of memory trace are ideally "taught" early on.

The young human COE will also learn thru experience with the random COEs in the environment. If these experiences result in a useful memory trace without impacting the homeostatic balance or metabolic processes they are labeled as "good experiences." Conversely if those experiences result in damage on the physical or even imaginary level the information processing which coordinates and arranges memory traces may be skewed sometimes to a permanent degree. What this means is the human COE will have an inaccurate "picture of reality" on the imaginary level. That inaccuracy will effect reactions that are not logically matched to perceived outside energy constructs. For example if a young human COE happens to eat an apple which is not ripe it may effect a homeostatic imbalance as the apple is being processed for nutrient extraction. The memory trace established on an imaginary plane may be one to not attract apples for nutrient extraction. This is an inaccurate premise on a logical plane in material reality because all apples are not less-than-ripe enough to eat. If by chance the memory traces associated with degree of ripeness are arranged at a later date the issue is put into perspective. If not the human COE will be ignorant with respect to apples permanently and have an none-attraction property about apples for no reason.

It is for this reason intentional education is the first choice with true facts in reality. It is also for this reason it is said that experience is a cruel teacher with higher costs.

Preliminary subject areas like the interpretation of symbols and use of symbols like words to effect energy transfer and transposition may take priority in education but are by no means the only areas where learning takes place. In fact almost every COE and active energy state, real or on an imaginary plane, has a compendium of stored symbols and images in light and soundwave form. This is termed subject knowledge or subject matter

As all education is high on the human COE priority list so is the systematized organization of knowledge stored for any specific COE that can be defined or labeled. The word *encyclopedia* stems from old Latin meaning "all around education." In human reality encyclopedic knowledge is as common as leaves on trees. In fact human COEs require specialized individuals or groups specifically labeled as to their memory trace arrangement with respect to a certain branch of knowledge.

For example a bicycle sales center needs to be staffed with human COEs who have an easily accessible memory trace arrangement with respect to bicycles. These traces could be images of individual material COEs which serve as constructs for a bicycle if assimilated in real, tangible, matter. The trace could also be a symbol as the quantity of currency required to exchange for a new bicycle. It could also be an arrangement of physical bicycle sizes in symbolic form that would match the size of the bicycle to the human COE who needs to attract a bicycle. The point is the human COE bicycle salesman has a unique set of memory arrangements which are all keyed on *bicycles*. That arrangement is the product of education but is also the engine producing education (teaching) for the buyer of the bicycle. That's why education is defined as *the transfer of memory trace* in human reality.

The reason humans find it advantageous to educate by subject specific groups is that the sum of all knowledge within the universe of human reality both on real and imaginary planes is so great in quantity that it is not possible for storage in one human COE's memory banks. Even if it were possible the time factor to process the information out and communicate it, sometimes across great distances might take years. At that much of a delay the end purpose of the desired knowledge will be useless because the window of opportunity will have passed for the desired energy transfer or transposition powered by information.

By having literally millions of specialty memory trace arrangements accessible at any specific point in time things can get done efficiently. The communication of knowledge by energy transfer is one of the most time consuming behaviors in human reality. If individuals are specialized in memory trace time is greatly reduced.

Recreation is defined as a "renewing." Renewing literally means to make new in this sense. So when human COEs are engaged in recreation they are trying to make something new. This is another aspect of human reality that operates on both an imaginary plane and within tangible matter. The tangible matter being the human COE. This is so because the thing they are probably trying to make new is the human COE himself. It's called a figure of speech but because human reality actually functions a good deal of the time on an imaginary plane recreation is taken seriously at face value of the symbolic word.

The experience of being "renewed" is essentially an outgrowth of the normal state of homeostatic balance. The difference being an added condition of temporary *euphoria*. Euphoria is defined as a state of happiness or joy. If you look up happiness its defined as joy or being content, satisfied, etc.. Essentially euphoria is a pronounced experience of homeostatic balance. This pronounced state creates the emotional illusion of "starting over" because the present end point in memory trace arrangement (thought) is obscured for a new emotional state. By virtue of normal memory most human COEs can recall a time of calm, without pressing needs or needs to attract something in order to be at peace. It is from that halcyon starting point that it is presumed eventual problems, responsibilities and needs began to cloud consciousness. This is a quasi-delusion but none-the-less the actual feelings of emotional and physiological balance allude to that imagined 'before" state.

The actual reason the human COE feels "renewed" is due to a number of factors. First the recreational activity serves to disconnect active memory banks from real-time priorities and needs necessary for the maintenance of metabolic processes as well as imagined necessities. Prior to that recreational activity memory traces are interacted on a regular basis until the goals are achieved. This is insurance that the memory trace will not dissimilate into oblivion so the proposed future action never takes place. If this happens the needs both central to the human COE or members of a group will never be met. At this level of concern it is a combination of both instinctual alarm and a learned manipulation of memory to achieve that end. The reality in most cases is that at that level of priority the memory is not easily dissipated.

A self-imposed distraction (recreation) serves to give those memory constructs an opportunity to replenish neuro-chemical energy reserves which are being diminished by constant conscious awareness of the need.

A second reason recreation serves to enhance emotional balance is that it often includes physical neuromuscular activity which ultimately releases endorphins and enkephalins which create a chemical euphoria. This is a natural biological, genetic-inspired mechanism which has ensured evolutionary survival. The exercise of neuromuscular energy is generally necessary in the acquisition of nutrients as well as defense. Exercise also replenishes cellular nutrients at an accelerated rate.

A third reason has to do with creative abilities. If the human COE can manipulate both physical matter and ideas on the imaginary plane he gains knowledge. That knowledge may enhance his ability to attract nutrients, arrange protection against rogue energy streams or aid in his defense against attack. This creative or inventive spirit has propelled the human COE up the ladder of civilization (which he has defined as such) since the beginning of his existence. This is fueled by instinct, much as a beaver builds a dam, as well as a learned behavior and ultimately the manipulation of memory trace on an imaginary plane.

The fourth reason is relief from boredom. Boredom means to become weary or tired. The word *tire* has an old English origin which means "a failure to come to an end." If a human COE is locked into a regimented response pattern to external energies as a method of maintaining metabolic processes and those external energies as well as the required response reactions do not change, the illusion of time not passing may be created.

This happens because of the basic definition of time in reality. Time is the transition of energy from either one state to another or the transposing of energy constructs. If there is an imperceptible change in whatever the time measuring stick happens to be it appears that no change has taken place, hence time has not passed. The sameness in external energy and required response creates a memory trace dictating a condition which "fails to end." This is labeled as *tiring* which also has come to mean fatiguing. Physical depletion of energy supplies for neuromuscular action results in fatigue and this happens if too much repetitive action takes place.

Fatigue generally has a negative effect on homeostatic balance. Even though much of this process only exists on an imaginary plane, for the majority of human COEs boredom from repetition of behavior does create a mild condition of imbalance. Recreation changes the behavior out of the repetitive pattern and restores the balance.

Entertainment is closely related to recreation in use, mechanics and effect. Entertainment is defined as amusement among other things. The word *amusement* is centered around an old Latin derivation as in comedic and humor. Humors were in essence "moistures" or liquids which when referenced to human physiology way back then were thought to be the responsible substances for feeling physically good (healthy) or bad (sick or infirm). Therefore the ideas of humor or humorous were applied to a condition of wellbeing or a feeling of wellbeing. That feeling of wellbeing was in fact what is labeled today as homeostatic balance. So if someone was in good humor they were medically sound and generally felt good. The use of the word crossed over into the performing arts as amusement or entertainment which was idealized as something to make people feel good.

Unlike recreation, entertainment does not generally require active participation other than observation. The many forms of entertainment include theater, music, dance and comedy.

Theater is an imitation of life by actors. Human COEs maintain a fascination with the observation of other human COEs in natural and staged settings and this is primarily instinctual. Some of the lower forms including primates maintain an imitative stance when observing other primates in behavioral settings and prefer to observe desirable situations like an ape feasting on bananas. Human COEs seek to create a memory trace from observations that may serve as transferable knowledge for future opportunities. Much of this is on an imaginary plane but enhances homeostatic balance just the same.

Music and the appreciation of music is unique to humans. There is a regularity of harmony that creates a predictable pattern. The human brain biologically favors the recognition of patterns thru light and sound perception. This was probably a desired feature supported by evolution because there is safety in predictable repetition. Safety translates into sustained maintenance of metabolic processes as well as homeostatic balance. Drugs like marijuana further drive this point home because they intensify auditory perception thereby creating an even more positive neural response to patterned auditory stimulation. The rhyme or reason behind the soundwave arrangement is immaterial as long as there is a predictable harmony and progression.

Dance is both a recreational activity and a form of entertainment. With respect to entertainment dance has elements of both theater and music. It is a predictable pattern of energy in motion and this ability is prized by the human COE. From early remnants of evolutionary necessity in both the hunt and the evasion and defense of the predator quick, graceful mobility is the key.

Comedy is unique in that it uses higher forms of logic within human memory constructs to evoke a pleasurable response. There are many theories as to why something is funny but it generally comes down to a state of tension or logical imbalance which is "rescued" very simply by a unique cognitive manipulation of memory. The surprise is the reward or return to homeostatic balance that was initially disturbed.

There is never anything serious about comedy in the normal sense but a mentally ill person might find something delightful that the general population does not. Mental illness is to be addressed later. Most comedy takes place in either a mode of theatrics or as a written or auditory account of real life events.

Any of the subject areas which come under the heading of entertainment achieve the same basic goals: First and foremost is a distraction from regular interaction with energy and standardized, repetitive responses (boredom). Second there is a quasi-learning feature in which there is either a logical construct of real matter or energy on an imaginary plane (the manipulation of ideas) which is new to the human COE and can be recorded in memory for possible future use.

Leisure is defined as free time. It is from old Latin meaning "to be allowed." The value in leisure is push-back from the concept of slavery or bondage as previously discussed. As is often the case the human COE will dramatize a simplified condition of energy to create a better quality entity on the imaginary plane. Even if there is no actual slavery or bondage issue the freedom from it can be imagined to create an enhanced homeostatic balance.

The human COE is considered at leisure when no mandatory or necessary pre-learned or rehearsed responses to external energy constructs are required and a slowed or "relaxed" option for neuromuscular or conscious manipulation of memory is totally subjective without consequence. The nuts-and-bolts mechanics

92

behind leisure are similar to the sleep state in that molecular-cellular replenishment and rearrangement of energy within the COE is achieved.

Although leisure is not high in the behavioral priority list for the human COE it is none-the-less an important, necessary requirement for both the indirect maintaining of an extended metabolic process set (extending life) and for maintaining a homeostatic balance while this is being achieved.

Twenty four hours to go
I wanna be sedated
Nothing to do, no where to go o,
I wanna be sedated

Just get me to the airport, put me on a plane
Hurry hurry, before I go insane
I can't control my fingers, I can't control my
brain
Oh no oh oh oh oh

The Ramones

Chapter Fifteen

Mental Illness

Human reality as stated before is pretty much determined by the *self*. If all of the selves have the same perceptual abilities and the same capabilities of memory and the same skills to manipulate those memories and the same neuromuscular mechanics to effect the transition and transposing of external energy accordingly we can conclude reality is the same for that entire group of "selves." The real life situation as it actually exists however falls far short. One of the biggest stumbling blocks to not only understanding human reality but to actually living in a functional human reality is the lack of understanding of the basic truism of how we define crazy. To cope with this aspect of life the administrative, governing, educated human COE group has come up with measuring sticks and labels to create a number of sub-groups which are presumed to be different from the main group of non-crazies. See below:

Words and Definitions: MENTAL ILLNESS: is defined as having any one or more of these disorders:

mood disorders (such as depression or bipolar disorder)

anxiety disorders.

personality disorders.

psychotic disorders (such as schizophrenia)

eating disorders.

trauma-related disorders

(such as post-traumatic stress disorder)

substance abuse disorders.

The word *disorder* is defined as confusion or chaos, lacking a normal condition of system.

The key to understanding all of this however is the word *normal*. Normal is derived from old Latin *norma* meaning of all things "a carpenter's square." If you don't know what a carpenter's square is, it's a piece of wood or metal with two legs similar to a standard ruler and those two legs intersect at exactly a 90 degree right-angle in the shape of an 'L'. It is used as a guide and a test-tool to make sure whatever was being constructed ended up as a straight and parallel assembly not prone to being skewed or crooked.

So from square-one (no pun intended) man has defined "normal" as having an arrangement which proceeded in a north-south or east-west direction without deviation. Pretty clever except if you consider this was pretty much pulled out of a hat without real support or reason other than he wanted his buildings to be level. If buildings were standardized as level and free from rogue angularity they were predictable with respect to form, function and integration. This straightness or "rightness" is a throwback to the concepts of quality, correctness, straightness and even law. The predictability contributing to the virtue of these features is driven by that same innate human COE desire for patterns and predictable repetition which contribute to a homeostatic balance. If an arrow is straight and follows a straight path it has a much better chance of hitting the target. If the arrow was curved it path would be "crazy."

All of the aforementioned classes consider mental illness a deviation from normal, predictable human COE behavior. All of the aforementioned also consider a rogue deviation in perception of external energies, memory encoding of external energies, manipulation of energies in the form of memory trace and action to control and move energy thru neuromuscular channels to be the cause(s) of mental illness.

Individuals designated as mentally ill are somewhat segregated from the larger group to different degrees based on the severity of the disorder. Some may be partially segregated pending the time of disorder onset and "cure" and others at a great physical distance for the life of the individual. The degree of segregation is directly related to the degree of inability the individual has to transfer and transpose energy as required by the larger group on a regular basis. Also the inability to independently maintain metabolic processes by the individual themselves may determine the degree of segregation.

The aforementioned classifications can be more specifically referenced to deviated energy states as follows:

mood disorders (such as depression or bipolar disorder): These are characterized by either a pronounced lack of homeostatic balance or a temporal difference in homeostatic balance from enhanced-to-lacking. Although these conditions are sometimes attributed to an internal energy fluctuation due to external energy influence, it is not always the case. In the worst cases the internal fluctuations cannot be reasoned out and genetic based structural deviations at the bio-molecular level are thought to be responsible. Derived energy in the form of chemical compounds is often used as a method to re-establish an energy balance at the neurological level. The group-interactive problem here is a lack of synchronization of emotion with external energy circumstance. Emotional states in any individual are communicated thru light and sound to other group members. That communication influences both the emotional states in the other group members as well effecting behavioral and attitude changes. Emotional states which are not in cadence with reality produce inefficiency in energy transfer and transposition. Also the emotional state of the afflicted individual may be so pronounced that it prevents him from dealing accurately with external energies as required by him.

anxiety disorders: Anxiety disorders are characterized by fear and worry particularly of the future. If you look up *anxiety* it is defined by worry. If you look up the word *worry* it is defined by anxiety. The old Latin origin for the word anxiety is *angere* which means "to choke." The old German / English origin for the word worry is "strangle or grab by the throat and tear." It is certainly clear how our modern word *anger* is a direct descendent. The anxiety and worry states are clearly developed from someone being on the opposite end of someone in a state of anger. It is also interesting to note that the word *hunger* is of old German /English origin and has only changed slightly in spelling but pronunciation is basically the same. The similarity in the phonetics of *anger* and *hunger* can't be denied. The crossover from someone or something in dire need of food to choose to grab some prey by the throat and tear or choke is believable.

In any event someone in a state of anxiety or worry is experiencing a state of homeostatic imbalance labeled as fear or uneasiness about some future or ongoing event. The event itself may be vague or even unknown but some uncertainty about it exists. Some of the physiological symptoms of anxiety are difficulty in breathing and an increased heartbeat as if being threatened with bodily harm. Of course all of this is generated on an imaginary plane. The effects on homeostatic balance however are the same as if it was "real."

The reason this is labeled as a disorder is obviously because the threat does not exist. There are cases of normal anxiety however when an individual is faced with a real situation proposed to materialize by way of accurate information, but that is not the mental health disorder. The actual mental health condition results from an error in perception and/or manipulation of memory trace. Essentially the individual has created a delusion characterized by an impending threat. That threat has mobilized energy to effect a state of neuromuscular tension which is essentially stress. This condition generally puts the individual in an alarm state which supersedes and encroaches homeostatic balance.

Anxiety may be precipitated to an abnormal level by either "learning experiences" or bad experiences resulting in a hyperactive energy transfer within the neural network. Chemicals are sometimes used to sedate the individual.

The reason the larger group has called out this individual as being somewhat dysfunctional is because the required focus for energy transfer and transposition in the group has been compromised by the preoccupation of eliminating the unreal threat.

personality disorders: Personality is defined as "of a person or characteristics or qualities of a person." Disorders of personality do not rank high on the list of severity as far as mental illness is concerned and most of these require no mandatory segregation from the larger group of people. That is of course unless that disorder has culminated in a behavior which is outside of the law. In fact most of the incarcerated population does exhibit personality disorder syndrome but there are probably an equal number of free people with the disorder as well.

The *characteristics* of a human COE are a label as to that person's historical response pattern to external energies of both human and non-human COEs. Human COEs being *pattern-centric* or having an innate ability to recognize patterns and create a corresponding memory trace, have applied this unique talent to the observation of other human COEs. Essentially as human COEs thrive in the ambient energy matrix they respond to specific energies in predictable, specific ways. This is considered normality. If any particular COE responds in an abnormal way on a regular basis a pattern emerges which "characterizes" that person as abnormal. If the behavior is not severe enough to inhibit functioning, it is labeled a personality disorder.

For example if attracted to an item in a store the normal response to the light incidence reflected from the item would be to execute a neuromuscular soundwave inquiring the quantity of currency required to exchange for the item in order to mobilize that item out of the store. If the human COE had a personality disorder they might choose to hide the item in their pocket and "forget to exchange it for currency." After getting home and remembering the item was not paid for they might interpret the item as not worth the requested exchange rate thereby keeping the item and keeping silent.

In another instance a normal response to a question about fish caught while fishing might be two or three, pending how many were actually caught. An individual with a personality disorder might respond with "seven, but someone took five from them when they were not looking." Another disordered personality might never drink beer because twelve years ago someone they knew drank beer and got physically ill.

Some of the worst cases of personality disorder entail a force driving the individual to create situations in which other human COEs might give their time and attention for no other reason than giving the disordered person enjoyment for wasting it.

The dynamic behind these disorders of behavior, attitude and thought is usually attributed to prior abnormal interaction with external energies that have created corresponding memory traces. Those memory traces power abnormal responses to identical, similar or in some cases complete different external energy constructs. That history might also be compounded by a "defect" in the memory storage system either genetic or because of influences like drugs or toxins. The word defect is in quotes because the word defect is a self-ascribed label by humans if something is considered outside of the way the majority memory operates.

In any event these are generally not serious disorders that can only be identified thru repetitive interaction and observance across different situations.

psychotic disorders (such as schizophrenia): These disorders are considered the most extreme with segregation from the larger group almost always mandatory. Psychotics are characterized by a gross inability to react to outside energies in ways to benefit the larger group or even the individual psychotics themselves. Everything from perception to memory trace to neuromuscular action may be affected rendering the individual either unresponsive to outside energies or responsive in ways so illogical on the normal plane of logic that they require a unique set of behaviors from ancillary human COEs to maintain their metabolic processes.

Homeostatic balance is also a wildcard and may be discernable or partially maintained by therapeutic interaction and drugs. Behavior, attitude, thought and action may be so impaired that the individual can cause physical harm both to others and themselves. On some occasions the psychotic may be able to function under a "ghost" of normality while maintaining an alternate identity. That identity might be so far removed as a serial killer murdering for no logical reason at all.

The cause of these disorders is almost always an imbalance or dysfunction in the energy transfer and transposition at the molecular level within the neural network of memory trace resulting in gross errors in the manipulation of memory trace. Conscience at any level is hit-or-miss and frequently non-existent at all.

eating disorders: These are characterized by the over-ingesting of nutrients *or* the under-ingesting of nutrients extracted from the proximal energy matrix. This results in either excess metabolic COEs inefficiently placed within the human COE or a lack of energy replenishment in existing COE structures within the human COE that are necessary for the efficient maintenance of metabolic processes. In either case longevity is compromised. The homeostatic balance however is usually maintained and in most cases is the direct driving force behind both behaviors.

The reasons behind the disorders can be varied. In the majority of human COEs a lack of nutrients will result in a normal homeostatic imbalance initiating attraction and extraction of nutrients from ambient external COEs either living or non-living. Once attracted, extracted and ingested the energy constructs within the nutrients replenish at the molecular level where needed. This itself initiates an energy transposition which contributes to homeostatic balance. Balance is restored until the cycle repeats due to the assimilation and ultimate transfer of those nutrient energies into the external energy matrix of the proximal environment outside of the human COE defined perimeter.

In the case of abnormal over-ingestion the homeostatic balance is not contributed to in quantities enough to restore it even though actual energy obtained from the nutrients is sufficient to meet immediate needs. This could be due to structural differences at a molecular level within the neural network or due to a self-imposed or other-imposed alteration of memory trace. That alteration consists of creating a "new normal" level of energy reserve past what is actually required. This may be a delusion based on ignorance and maintained thru repetition as habit. The individual is very plainly addicted to food.

As ingested nutrient reserves are positioned within the human COE in greater amounts the internal energy constructs which structurally hold this reserve adapt by adding more constructs. With the subsequent absorbing of ingested nutrients past the required level more and more internal energy constructs are added spiraling the disorder to an even greater extent. The net result is a metabolic system which operates inefficiently due to blockages and delays in energy transition and a greater mechanical stress on neuromuscular systems for general mobility and form..

In the case of abnormal under-ingestion the homeostatic balance is not affected by a deficit in nutrient energy reserves and may in fact be compromised by the actual attraction, extraction, ingestion and assimilation processes normally utilized to maintain metabolic processes. The end result is the obvious lack of supporting energies creating inefficient metabolic operation and eventually a cessation of metabolic activities (death) if the pattern continues.

The driving forces behind abnormal under-ingestion are similar to the situation of over-ingestion. There may be a "defect" at the neuro-molecular level causing a lack of energy transference to disturb the homeostatic state when energy reserves are low and in need of replenishment. There may also be abnormal response patterns in perception linked to memory trace or accurate perception linked to errors in the memory trace system compounded by problems in memory trace manipulation. This may also be due to a similar neural network defect or can even be learned and maintained as a delusion of reference.

In the case of delusion the individual has a measuring stick on an imaginary plane which compares either real-time images or past memory trace images with respect to physical size and weight of normal human COEs with their own self-image. Some of those images are of a random average of the larger group and some of those images being of that particular individual (self-image). Due to a distortion in image transfer from real to neural energy construct, the comparison is skewed. In the under-ingestion disorder the comparison of images always results in the individual being physically larger and weighing more than the reference group's images. This is motivation to reduce nutrient intake. Once again this abnormal response can be learned or may have resulted from negative experiences with a correct nutrient ingestion schedule.

Eating disordered individuals are rarely segregated from the group unless they have reached a level where specialized care is required to maintain the continuance of metabolic processes. They are generally treated by re-calibrating the memory trace with homeostatic balance by specialized members of the larger group. In some cases drugs are used to balance energies in favor of normality.

trauma-related disorders (such as post-traumatic stress disorder): These disorders are basically due to real experiences in which external COEs human or not, have caused a homeostatic imbalance in the individual and that homeostatic imbalance is sustained even when the initial causing energies are not present.

The core of this disorder lies in a memory trace caused by the original, disturbing, outside energy force. That memory trace was not dissipated as other

101

unneeded memory traces are. The reason for this may lie in the level of transposition of the disturbing energy from external to internal-neural. The term "post-traumatic" refers specifically to this. After the initial disturbance the individual might unintentionally experience trauma related energy constructs on an imaginary plane. It could be light or sound energy or any combination. On cue with the unintentional recollection a homeostatic imbalance occurs similar to the original experience. This in itself is distressing causing an inefficiency in normal energy processing. It might also cause other ancillary conditions where normal perception is skewed to be sensitive to any energies similar to the traumatizing energies. This creates phobias or fear and avoidance of benign ambient energies in the environment. The problem once again is that activity on an imaginary plane is resulting in disordered control and effect of energy in the "real" world.

Treatment usually consists of the attempted dissemination of the rogue memory trace thru re-learning as well as drugs to restore homeostatic balance.

Individuals who have this disorder are generally not segregated from the larger group on the basis of the disorder alone but often choose to segregate themselves as a way to minimize episodes of imbalance.

substance abuse disorders: These disorders are the easiest to define. Essentially there are specific COEs which exist at the molecular level which when ingested either restore homeostatic balance that has been previously been disturbed for whatever reason or create a condition of exaggerated balance superseding normal circumstances. These COEs are labeled as psychoactive drugs or chemicals or substances.

They cause these changes to homeostatic balance by dissipating rogue energies once ingested. Those dissipated energies find their ways into the homeostatic alarm neural-network and block natural alarm energies. They do this by attracting molecules of alarm energy into bogus situations for dissipation and neutralization. The drugs or chemicals might also dissipate energies into controlling or normal energy regulating channels causing natural internal energy reserves to be released for transposition even though there is no need for them for the sustaining of metabolic processes.

The down side to all of this is the bogus state of homeostatic balance and energy reserve expenditure which may compromise normal transfer and transposition of energies as required for maximum longevity. Essentially it is a disorder of energy within the individual which puts the individual out of cadence with the ambient energy environment. Because the drug or substance mimics normal signals of "rightness" the individual may adhere to the delusion to the extent of continuing the ingestion with regularity. This is termed addiction.

The problem with addiction is two-fold. First there is the neglect of normal metabolic process maintenance because there are no alarm signals. Second is the compromise in normal energy transference, control and transposition as required by the larger social group.

Treatment generally involves some degree of relearning and abstinence. Sometimes it is successful and sometimes not. True substance abuse disorder is considered a serious problem by the larger social group and may have some degree of permanent affliction.

In conclusion mental illness is a sometimes transient, sometimes permanent state of human COE operation in which the "normal", anticipated, energy control, transfer and transposition within the ambient environment both for the benefit of the individual and the larger group is compromised by energies outside of the direct, immediate control of that individual. It is a natural part of reality in which human beings have adapted to human reality.

She blinded me with science
She blinded me with science
And hit me with technology

Thomas Dolby

Chapter Sixteen

Technological Advancement and Progress

Words and Definitions: TECHNOLOGY is defined as applying science for a practical purpose. *Science* is defined as the systematized collection of knowledge in the real world.

Therefore technology is the encoding into memory by symbols or otherwise perceivable differences in energy constructs and how they interact (facts). Those facts are then aligned on an imaginary plane to be effected in real-time towards some benefit of human COEs. The main benefits purportedly targeted are almost always geared towards the extension of metabolic processes (longevity) as measured against the normal transition of energy in the universe (time) and an increase in the amounts of homeostatic balance for the human COE during their lifetime.

The greater the amount of this technological process transfer (fact collection, manipulation on an imaginary plane and energy transposition in the real world), the greater amount of *advancement* is assumed to be reached.

The concept of advancement however is a human creation whereby the change in the graded nature of civilization is measured against time. Time as you recall is the normal transition and transfer of energy in the universe and everywhere else. So the human ideal is purported to create as many benefits for the extension of human longevity while maintaining as great amount as possible of homeostatic balance while enduring the least amount of universal energy transition (time). This is the most desirable condition labeled as *the marked advancement civilization.*

Human COE's generally experience some degree of homeostatic balance even when they manipulate memories on the subject of technological advancement on an imaginary plane.

If that isn't complicated enough, *Progress* is defined as advancement towards a more complete or modern state of affairs. The word *modern* is the key to the puzzle here. *Modern* is defined as referring to the present as opposed to the past. It originates from the early Latin *modo* meaning "just now."

So right out of the gate it's a paradox. Progress being an advancement towards the present *and not an advancement from the present into the future.* This is one reason why progress is often slow and arduous. Humans in essence are trying to advance to the point where they already are. If you are already there, no work is required but there is no advancement either. This sets up a condition of paradoxical incongruence.

Incongruence violates the innate human affinity for the concepts of purity, quality, straightness or "rightness." As human COEs struggle to make a logical advancement within an illogical frame of reference a state of tension or mild homeostatic disturbance is tolerated. If we simply reverse this chicken or the egg puzzle we have: *A mild state of homeostatic imbalance is experienced by the group as a norm and that imbalance is attributed to stagnation or a need for advancement.*

As long as the subject of advancement ranks in the upper- levels on the human COEs' list of priorities there is a temporary reprieve from that particular homeostatic imbalance. This however is like the moth chasing the flame because once something has materialized as an "advancement" or "progress has been made" the cycle starts all over again. That's why whatever the new coconut is; a smart phone, Alexa, self-driving cars or a cure for cancer it is never enough.

The majority of human COEs are unaware of this paradoxical arrangement of energy, even though it creates a bottomless pit of the need "to advance." The safe "advancement" escape is the belief in a more complete, efficient state of affairs which will create more homeostatic balance in a greater number of human COEs while extending their metabolic processes for greater length of time. These are the basic goals of the human COE since his genesis and are mostly hard-wired instinct. The idea, though noble, may only be instrumental towards another goal almost outside of the grasp of human consciousness.

If you strip away the self-aggrandizement and the innate drive for quality and "rightness" you are left with a shell of behavioral motivation which comes down to a desire to control energy. This is regularly peddled as man's conquest over nature and ignorance. Reality though indicates a different picture.

Those innate features necessary for survival positioned in noble "rightness" and virtue have been translated into energy control. That control over energy is instrumental in the re-creation of all aspects of living humanity in non-living energy constructs with nearly limitless function when measured against the capabilities of man. Essentially if energy can be controlled without restriction it

106

takes place the human COE will become the intellectual minority. Perhaps even maintained as a sample of the past like a living fossil.

Our father high in heaven, smile down upon
your son
Who's busy with his money games, his women
and his gun
Oh Jesus save me

And the unsung Western hero, killed an Indian
or three,
And then he made his name in Hollywood
To set the white man free
Oh Jesus save me

Jethro Tull

Chapter Seventeen

God and Religion

Words and Definitions: GOD: Originates from early German and from early Indo-European *GHU-To* which means *to call or to invoke.*

RELIGION: the belief in and worship of a superhuman controlling power, especially a personal God or gods. From old Latin, *to bind as in obligation or reverence.* Reverence is from old Latin *revere* which means to stand in awe. Awe means wonderment or astonishment but the old English origin means *terror or dread.* So religion and God generally defines a supernatural existence which is basically imperceptible under normal conditions of reality but has supreme power over everything on earth and beyond. That supreme power also commands respect.

There are over 4,000 religions world-wide and there have probably been many more since the creation of man. The oldest organized, documented religion is Hinduism dating back to 1,500 BC. It is safe to say there has probably been some form of religion since the beginning of humanity because wherever man went religion followed. This is another unique feature of the human COE which is not shared by the lower forms of life.

Although the generic nature of God and religion is on par with UFOs and ghosts because of supernatural elements, the belief in God has been infinitely more popular than the belief in UFOs, ghosts, etc..

Because scientists cannot empirically define God there also are those who deny existence as well as those who really don't know either way. Because the vast majority of humanity does have a belief in God today and has so since the dawn of man, it can be concluded that there must be something to it. If a caveman gave attention to anything at all you can bet it was absolutely pivotal for survival.

To understand God we need to start with the very same basics of human reality. Human reality consists of constructs of energy derived from perception and subsequent memory traces within the human COE as agreed to by the larger group of human COEs. If by some chance there was a contagious outbreak of permanent schizophrenia caused by whatever unknown reason and the vast majority of the population was seeing monsters in the trees and hearing choruses in some strange, unknown language, human civilization would be vastly different than it is today. Human reality might be shaped so that trees were shielded from view by law and perhaps headphones would be worn as a mandatory course of normal events. Anyone who did not follow those societal dictates would be thought of as crazy the same way as someone stripping off their clothes and going for a walk naked by today's standards.

Therefore in this hypothetical case of mass schizophrenia we have energy constructs which were actually either distorted thru perceptual energy transformation channels or in the transposing of the correct perception into a distorted memory trace or a combination of both. The end result is still an understanding of the external energy matrix on an imaginary plane. That's why reality within the scope of humanity exists for the most part on an imaginary plane even though it is responded to on a non-imaginary plane.

The paradox here is that once energy is effected thru some action or transference based on an imaginary plane the immediate feedback is a transposing of that action to perceptual energy with subsequent new memory trace creation. The net result is that whatever was "happened" thru that action is re-encrypted in memory as an interpretation of reality or "real." Even though it was the product of a distortion in the first place the human COE considers it accurate because he has no other measuring stick to test it. Distorted or not it is normal reality to him.

The reason the understanding of how reality is put together is relevant to the concept of God and religion is that God exists on the imaginary plane for the majority of human COEs. It's true there have been people who have attested to seeing and hearing God on the non-imaginary plane (the real, tangible world) but this has been a rare occurrence. Jesus walked the earth thousands of years ago and sightings of the Virgin Mary as well as probably a thousand other miracles have occurred across all religions but the vast majority of humanity has not experienced this. Since reality is determined by virtue of the majority human COE agreement it can be concluded that God does exist and God is real but for the most part on an imaginary plane.

This is not as extreme as it sounds considering every event that has ever happened or will happen only exists on an imaginary plane. The present, though non-imaginary is a brief moment in time which slips into the past as part of normality.

Words and Definitions: SUPERNATURAL: (of a manifestation or event) attributed to some force beyond scientific understanding or the laws of nature.

That's the official definition but from the perspective of pure physics it comes down to the interpretation of energy. A supernatural occurrence is some kind of perceivable energy construct capable of creating a memory trace in which the immediately prior transposition or prior transition of energy to effect that perception has never been established in a human memory trace. It is literally "not traceable."

If for example a flashlight were to come "on" without anyone being near it, it might be considered a supernatural event. If on the other hand further investigation found the switch was affected by ambient moisture or temperature it would not be considered supernatural because the precursor energy causing the event could be identified. In addition this same precursor energy had already been encrypted in memory banks as being capable of similar causative events.

If however the conditions of moisture and temperature were not found to have affected the switch the occurrence might be considered supernatural. If further investigation concluded the switch was absolutely sound and it was confirmed thru documentation and/or human witnesses that it was last touched and left in the "off" position, it could be concluded with certainty that a supernatural event did indeed occur. If this happened on more than one occasion it would be further validated.

At the end of the day it can be concluded that some form of energy acted on the switch causing it to move from *off* to *on* but that energy has left no clue as to its identity and it does not exist in any human memory trace.

Events similar to this example have occurred throughout history some being documented and sworn to. Some part of the population does believe these to be real and others are skeptical because of the lack of repetition for most events. Also credible recordings by camera or audio are scarce, if at all. This is particularly true in the case of ghost or alien sightings even though alien abductions have been sworn to. UFO photos are also speculative because none of these vehicles have ever landed long enough or left a tangible piece of matter as evidence of their existence beyond the imaginary plane. Unlike the case for God, attributed changes in energy arrangements are sparse and weakly supported.

It can be concluded the vast majority of the scientifically astute, modern civilization does not believe supernatural occurrences happen but are more inclined to suspect ignorance, miscommunication or an error in memory or perception to be the driving forces.

In contrast a belief in God and adherence to some kind of religion is markedly higher across humanity. God and religion are also widely promoted via the air waves, educational institutions, the press and ultimately religious organizations. Family groups reinforce this much of the time in normal child development because it usually has a positive effect in attitude and behavior for the later adult.

With respect to God it appears that there is something happening on an imperceptible level without an identifiable energy trail which effects energy transfer and transposition on a perceptible, tangible plane within the human sphere of reality.

The mechanics behind religion might give some insight into the subject and its vast popularity. Firstly most religions have some things in common. First and foremost is a belief in a deity or deities or deity-like force. That deity is always omnipotent and can perform any conceivable task at whim. That deity is usually thought responsible for the creation of all perceivable existence both in living and dead tangible matter and energy forces such as the sun's rays. That deity is also man-centric in that humanity is his creation and human behavior is his observation and focus. He also most certainly has a policy to extend existence past the cessation of metabolic processes for the human "spirit."

Spirit is defined as *the nonphysical part of a person which is the seat of emotions and character; the soul.* The word spirit is from old Latin meaning breath or to breathe. If you look up the word *soul* it just says *spirit.*

All organized religions have a documented set of laws the deity has specified as a guide for human behavior. If the laws are followed humans are rewarded, if not they are punished. The laws are generally coordinated along lines of logic which will either extend metabolic processes to their maximum or ensure a homeostatic balance not only for the individual but for the group as well. The added bonus for obeying the laws is the secured extension of the spirit into some other form of utopian homeostatic balance after the cessation of metabolic processes. Failure to adhere to laws while living is thought to cast the spirit into a lack of homeostatic balance on a temporary or permanent basis.

Most religions also have ceremony-specific actions and behaviors which have been determined by the original adherents as a way to better follow the deity's dictates.

The communicative threshold between the deity and humanity in most religions can be crossed by way of prayer or chant. Prayer is defined as a humble request. The result of prayer throughout history has much of the time been a request granted. This is one reason why the belief that God exists is so widespread.

God which no *living* person has ever seen and which is not definable other than an omnipotent force without beginning or end can be tentatively labeled as some form universal energy. This is not a new idea as philosophers like Descartes, Bruno and Spinoza have alluded to the same premise centuries ago, sometimes being persecuted as heretics for such beliefs. The ideas that all things matter are made up of a same basic substance was put forth surprisingly by people who knew nothing of atomic physics or the concept of basic energy is amazing. Also an interconnection of all matter and thought was contemplated as some kind of connection with supreme control.

Although a number of human COEs have testified to coming back to the material world after death that end of belief must be left up to speculation and faith. The problem here lies in the accepted definition of death. Generally speaking a cessation in cardiac activity and lack of brain waves is the present day determination of death. Because some examples of life "returned" have passed thru those points it may be understood as life after death but the actual validity of those points needs to be examined. Because the essence of life in any living thing is active energy there may be a pool of that energy nested in the protoplasm remote from the heart or brain. If a patient is comatose he may indeed be "brain dead" as far as cognitive functions are concerned but sometimes cognitive abilities are regained. A similar occurrence may take place in the case of those who have "returned."

As for differences in the actual organization of different religions it might be concluded that there is a credible reason for those features and the differences may have evolved just as differences in language and culture. Also differences in the specifics of evidentiary enlightenment during the infancy of the various religions via visitations, miracles, etc. may have shaped organizational themes.

The biggest factor to consider in the sustained need for religion is that it has proven a positive force in both the extension of human COE metabolic processes as well as the maintenance of homeostatic balance. It has also served to shape civilization into being more "civilized." Religious laws are often similar to civil laws and this serves to reinforce behavior in a guided direction. The argument against this is the history of persecution and war that has taken place over differences in religious ideology.

From a pure physics perspective the analysis of why religion and prayer works might come down to that same energy transfer and transposition that controls everything else. Firstly God is not a genie in a lamp no matter which deity is your deity. If you pray to God for a new watch because you just want a new watch, he will probably ignore you as all religions teach. But if you have a genuine need which is motivated by either imminent danger to metabolic processes or something that causes a significant homeostatic imbalance, there is a good chance the situation will be rectified. This is particularly true of situations in which an individual human COE is mal-aligned with the larger group or not in cadence with rogue energy which presents itself as a stumbling block.

The stipulation of humility in prayer is evidenced as a precursor to a favorable response from God. It is related to the importance of the request and amount and sincerity of the prayer. Organized religion enables groups to pray for any particular need and this emphasizes the condition of need. What is actually happening here is a human COE in a state of homeostatic imbalance due to either a direct metabolic alarm, an indirect metabolic alarm (a warning of pending physical danger) or an alarm on an imaginary plane is sensed by the group. Even though the source may be imaginary due to ignorance, disordered perception or inaccurate manipulation of memory trace the resulting imbalance is just as real. The imbalance in all cases causes the human COE to effect prayer either by neuromuscular sound or cognitive recollection by way of the manipulation of memory traces. What is happening is an energy arrangement initiated by the imbalance is being transposed thru a neural network into cognition and neuromuscular energy further transposing external ambient energies by sound waves and brain waves.

If reality is made up of energy constructs and God is a force of universal energy a detectable condition of negative energy is creating a blip in the big picture. It however is not just any blip as you might generate by commercial advertising begging for customers. This blip carries with it a requisite for energy arrangement which will change it from negative to positive.

Negative and positive are interchangeable terms. The important thing is the polarity difference. This is important because all energy at its core is basically a flow across polarity from *something* to *nothing* or *nothing* to *something.*

The reason prayer needs to be a humble request is because "humble" gives it a directional trail as coming from a "needing" energy source without control of that particular energy arrangement causing distress. Humble also connotes sincerity as in genuine. When it comes to energy transposition there is no speculation or choice. It's like nature at face value. Humble also creates an open channel for a flexible response with which there will be no repercussions as to quality, purity, completeness or "rightness" eliminating the potential for another negative blip. In essence the "humble" aspect tags a positive aspect to a negative energy position.

It might be noted that commercial advertising as well as big business does try to cross the threshold from one of profit generation to one of sincerity in hopes of a positive energy change in their direction

It might be argued that if God is indeed a universal energy force flowing infinitely over all energy in the universe including human COEs and that universal energy has the capacity to influence or "control" all other energy transfer and transposition it still presents a puzzle as to how God is able to "hear" requests and execute change. This is not a difficult puzzle if you recall that everything in human reality is made up of energy including words and symbols.

If a farmer in Thailand is in desperate need of rain and prays for it and a farmer in the U.S. does the same, they are reaching out in two different languages. Although the languages are different rain is rain and it is the same wet COE in molecular form needed to be transposed *from* other energies so that it can be attracted to the earth for transposition *into* still other energies to effect crop growth. The keys here are the farmers' positions in the energy matrix along with the COE rain itself, and the wave-form symbol for rain, the actual word "rain." When that word was coined as the symbol for falling water molecules that symbol immediately became part of the energy matrix. Just because man created a symbol for rain as an audio wave-form-word does not mean it precludes membership and effect in the ambient energy matrix.

When those farmers came into existence they too became part of that same matrix. To be more accurate they became part of the energy matrix in the form of human COEs *in proximity* to land, seeds, crops, rain, etc. which all had particular levels of attraction to each other and potentials for interaction. The precursors to the farmers, crops, etc. were other forms of energy that had been transposed as seeds and land which is a composite of dissimilated plants and other nutrients. What we now label as the farmers, land, seeds, crops, etc. are actually energy arrangements within a slice of time. A veritable snap-shot of energy in transition.

The net result is the same as the genesis of matter and non-matter from square-one: Energy is doing an intermingling dance in which there is only abundance and deficit trading places constantly.

The praying farmers are active COEs indicating a deficit of rain needed to be filled within the energy matrix to balance the other COEs. Because time is a man-made concept it is hard to grasp that all of these events are happening at the same time and not really in serial as it appears. The deficit creates a homeostatic imbalance within the universal energy system and the prayers are an energy connection to voice that imbalance.

Sometimes however prayers are not answered at all because the request does not fit well into the bigger energy matrix of events. More often other times prayers are circumvented by new circumstances which satisfy the homeostatic balance indirectly. That's why they say "God moves in strange ways or God draws a straight line but does it with a crooked pen."

As far as modern physics is concerned the essence of universal energy can be labeled God. God being responsible for the creation of anything is a mix of theoretical particle physics and "proven" components like the Higgs Boson or even Einstein's transposition of matter to energy and the reverse. The problem however

116

is putting God in an intellectual framework with these methods. It may not be possible within the life-span of man as a living species on earth. Reasons as follows:

First and foremost is a set of cascading quantum-like theories which base probabilities on other probabilities. Physicists hope super-computing will one day number crunch enough probabilities to gain some kind of useful understanding of the science so it can be translated into technology. The other problem is the micromanaging of extremely small, almost imperceptible energies, or particles of energy with accelerators to prove they exist. In any case the apparatus necessary to get to any tangible point of significance on the human level would not be practical for perhaps thousands of years.

When the technology is that far out the impact of the earth's longevity and subsequent human existence enter the equation. The abbreviated longevity of both may compromise any meaningful progress. This in fact may be one of the driving factors to get the control of energy consolidated on a non-living plane, perhaps somewhere safe out in space where a continued intellect may survive for hundreds of thousands of years..

It makes more practical sense in immediate human reality to encapsulate energy's reign from one-step back as a conglomeration or a consolidated universal matrix of which man is a small part. How it may have reached that particular arrangement and when is purely speculative without supporting evidence at this point in time. Following that reasoning it makes similarly more logical sense to consider divinity based on fact rather than conjecture or stipulation. Once again as stated early on, if a tree falls in the woods but no one is there to hear it, does it make a sound? Physicists are pressing the argument both ways but as far as the scope of human reality is concerned we have to say it does. By the same reasoning the testament of millions of human beings who claim they has experienced divine intervention as well as thousands of documented miracles and visitations push the scale towards an existence of God in the real world.

Out of the way
It's a busy day
I've got things on my mind

Pink Floyd

Chapter Eighteen

Human Reality In Retrospect

To sum it up: Based on proven science nested for the most part in the subject areas of physics, and documented, historic human behavior and thru thought patterns defined as normal, human reality is a direct subset of the behavior and interaction of energy.

Although man may claim to be the author of civilization, technological advancement and even compassionate behavior (which he has so nobly labeled as "humane") he is merely an energy shell in the matrix of countless other energy influences.

Though he prides himself for the ability of conscious decision based on ideas and memories, those options and seemingly subsequent consequences were most probably already mapped out in perfect cadence with the "big picture." An energy matrix he cannot escape from.

Positioning lower forms of life on his own self-created measuring stick he was able to claim victory and supremacy. The words *victory* and *supremacy* were also man's creation. For now the human COE can command the ship within limits of nature and he really has no choice to do otherwise. He may consider it his option to save the whales or curtail global warming but whichever way he turns the energy matrix will adapt. Even if to one extreme is the threat of a nuclear wasteland there will always be those who extend life further as the earth continues to spin.

Where is man headed? As man makes his way across time on his high horse he might just be heading towards a consolidation of human consciousness in non-human, non-living media like silicon chips or atoms of copper. After all that very consciousness he claims his own is an arrangement of energy masked with goals created by man himself. If man is taken out of the equation you are left with rogue energy. Does rogue energy have a goal? Does rogue energy even have purpose to need a goal? These are questions beyond the imaginary plane of human capability. When energy took the form of man it had no idea existence could be so complicated.

Black
And blue
And who knows which is which and who is who
Up
And down
And in the end it's only round 'n round
Haven't you heard it's a battle of words

Pink Floyd

Chapter Nineteen

A Cosmic Shade of Jade

The title of the book really doesn't have anything to do with anything. It's just an interesting arrangement of word symbols. You may have construed some deeper meaning to the phrase because that's one peculiarity of man which at some point creates a component part of the human reality he lives in.

A pattern of symbols of energy on an imaginary plane, manipulated by preformed energy in the shape of a memory trace. Whether accurate or not perhaps a new conglomeration of energy will be effected thru human action on a non-imaginary plane and a new color will be born. Defined, labeled, categorized and organized at some position within the spectrum. A new construct in human reality that will add to the extent of man's understanding of the universe. But at the end of the day it's just a microcosm of the position and arrangement of energy in the universe consciously interpreted as a part of human reality for no other reason than *that's the way it happened.*

As far as the book itself is concerned, it is just a collection energy in the form of words which symbolize other energies, their transpositions, interactions, arrangements and rearrangements on a non-stop basis. If it had been a book about baseball or trees or fishing, those symbols would have been confined to active energies specific to those constructs of matter and actions characteristic of those constructs. Because the subject of the book is human reality the energies symbolized by words are a human COE's conceptualization of all energies external to a memory trace created within his mind and his response in the form of other effected energies within his control. In effect the book is a cross-section or slice of the organization of energy within man's daily existence both intrinsic to man and the external universe. Of course no one actually interprets life outside of the "normal man-made drama" but looking forward perhaps it should be. interpreted in light of pure physics. It may help to cure the world's ills like war, violence and greed, but that's another can of worms all together. As far as posterity is concerned the book alone is little more than a chimpanzee looking at his reflection in a mirror. A glimpse of reality but none-the-less an illusion.

If you enjoyed this book also by James Ciccone available at Barnes and Noble or online at Amazon:

The Mild Equator

Elephants In Real Time

www.ingramcontent.com/pod-product-compliance
Lightning Source LLC
Chambersburg PA
CBHW030806180526
45163CB00003B/1165